T0200734

# CARNAP, QUINE, AND PUTNAM ON METHODS OF INQUIRY

Carnap, Quine, and Putnam held that in our pursuit of truth we can do no better than to start in the middle, relying on already-established beliefs and inferences and applying our best methods for re-evaluating particular beliefs and inferences and arriving at new ones. In this collection of essays, Gary Ebbs interprets these thinkers' methodological views in the light of their own philosophical commitments, and in the process refutes some widespread misunderstandings of their views, reveals the real strengths of their arguments, and exposes a number of problems that they face. To solve these problems, in many of the essays Ebbs also develops new philosophical approaches, including new theories of logical truth, language use, reference and truth, truth by convention, realism, trans-theoretical terms, agreement and disagreement, radical belief revision, and contextually a priori statements. His essays will be valuable for a wide range of readers in analytic philosophy.

GARY EBBS is Professor of Philosophy at Indiana University, Bloomington. He is the author of *Rule-Following and Realism* (1997) and *Truth and Words* (2009) and coauthor of *Debating Self-Knowledge* (Cambridge University Press, 2012). He has also published articles on a wide range of topics in the philosophy of language, logic, and mind, as well as epistemology and the history of analytic philosophy.

# CARNAP, QUINE, AND PUTNAM ON METHODS OF INQUIRY

GARY EBBS

*Indiana University, Bloomington*

CAMBRIDGE
UNIVERSITY PRESS

# CAMBRIDGE
## UNIVERSITY PRESS

University Printing House, Cambridge CB2 8BS, United Kingdom

One Liberty Plaza, 20th Floor, New York, NY 10006, USA

477 Williamstown Road, Port Melbourne, VIC 3207, Australia

4843/24, 2nd Floor, Ansari Road, Daryaganj, Delhi – 110002, India

79 Anson Road, #06-04/06, Singapore 079906

Cambridge University Press is part of the University of Cambridge.

It furthers the University's mission by disseminating knowledge in the pursuit of education, learning, and research at the highest international levels of excellence.

www.cambridge.org
Information on this title: www.cambridge.org/9781107178151
DOI: 10.1017/9781316823392

First published 2017

Printed in the United Kingdom by Clays, St Ives plc

*A catalog record for this publication is available from the British Library.*

*Library of Congress Cataloging-in-Publication Data*
Names: Ebbs, Gary, author.
Title: Carnap, Quine, and Putnam on methods of inquiry / Gary Ebbs, Indiana University.
Description: New York: Cambridge University Press, 2017. |
Includes bibliographical references and index.
Identifiers: LCCN 2017004160 | ISBN 9781107178151 (hardback)
Subjects: LCSH: Methodology. | Hermeneutics. | Carnap, Rudolf, 1891–1970. |
Quine, W. V. (Willard Van Orman) | Putnam, Hilary.
Classification: LCC BD241.E224 2017 | DDC 121–dc23
LC record available at https://lccn.loc.gov/2017004160

ISBN 978-1-107-17815-1 Hardback

*For Martha*

Our own discipline, logic or the logic of science, is in the process of cutting itself loose from philosophy and becoming a properly scientific field, where all work is done according to strict scientific methods and not by means of "higher" or "deeper" insights.

Rudolf Carnap, "The Task of the Logic of Science"

Unlike Descartes, we own and use our beliefs of the moment, even in the midst of philosophizing, until by what is vaguely called scientific method we change them here and there for the better. Within our own total evolving doctrine, we can judge truth as earnestly and absolutely as can be; subject to correction, but that goes without saying.

W. V. Quine, *Word and Object*

The difference between statements that can be overthrown by merely conceiving of suitable experiments and statements that can be overthrown only by conceiving of whole new theoretical structures – sometimes structures, like Relativity and Quantum Mechanics, that change our whole way of reasoning about nature – is of logical and methodological significance, and not just of psychological interest.

Hilary Putnam, "It Ain't Necessarily So"

# Contents

*Acknowledgments*                                                    *page* ix

Introduction                                                              1

PART I   CARNAP

1   Carnap's Logical Syntax                                              13
2   Carnap on Ontology                                                   33

PART II   CARNAP AND QUINE

3   Carnap and Quine on Truth by Convention                             57
4   Quine's Naturalistic Explication of Carnap's Logic of Science       95

PART III   QUINE

5   Quine Gets the Last Word                                           113
6   Reading Quine's Claim That Definitional Abbreviations
    Create Synonymies                                                  128
7   Can First-Order Logical Truth Be Defined in Purely
    Extensional Terms?                                                 144
8   Reading Quine's Claim That No Statement Is Immune
    to Revision                                                        168

viii                          *Contents*

PART IV    QUINE AND PUTNAM

9    Conditionalization and Conceptual Change: Chalmers in
     Defense of a Dogma                                          191

10   Truth and Transtheoretical Terms                            205


PART V    PUTNAM

11   Putnam and the Contextually A Priori                        233

*Afterword*                                                      257
*Bibliography*                                                   261
*Index*                                                          273

# Acknowledgments

For helpful comments on previous drafts of the essays in this volume I am indebted to many friends and colleagues. These include, for Chapter 1, Stewart Candlish, Michael Friedman, Richard Gaskin, Tim McCarthy, Sanford Shieh, and Joan Weiner; for Chapter 2, Kate Abramson, Richard Creath, Vera Flocke, Gregory Lavers, Joan Weiner, Stephen Yablo; for Chapter 3, Kate Abramson, Susan Blake, Matt Carlson, André Carus, Jim Conant, Richard Creath, Dagfinn Follesdal, Greg Frost-Arnold, Warren Goldfarb, Steven Gross, Peter Hylton, Mike Koss, Penelope Maddy, Tim McCarthy, David McCarty, Charles Parsons, James Pearson, Chris Pincock, Huw Price, Georges Rey, Tom Ricketts, Dennis Senchuk, Sanford Shieh, William Taschek, Steve Wagner, Meredith Williams, Michael Williams, the editor of *Mind*, and three anonymous referees – two of whom have subsequently been identified to me as Marian David and Eric Loomis; for Chapter 4, Michael Friedman, David Macarthur, Sean Morris, and Tyke Nunez; for Chapter 5, Kate Abramson, Susan Blake, Matt Carlson, Michael Friedman, Jim Higginbotham, Adam Leite, Janet Levin, Mark Kaplan, Mike Koss, Andrew McAninch, Blakely Phillips, Tom Ricketts, Barry Stroud, Steve Wagner, and Joan Weiner; for Chapter 6, Lieven Decock, Hans-Johann Glock, Catarina Dutilh Novaes, Jeanne Peijnenburg, Stewart Shapiro, Sander Verhaegh, and Joan Weiner; for Chapter 7, Sarah Adams, Micheal Beaney, Dylan Black, Sinan Dogramaci, Mike Dunn, Kirk Ludwig, Chris Pincock, William Taschek, Joan Weiner, and an anonymous referee for *The British Journal for the History of Philosophy*; for Chapter 8, Yemima Ben-Menahem, Frederique Janssen-Lauret, Jeremy Heis, Peter Hylton, Mark Kaplan, Kirk Ludwig, Andrew Lugg, Hilary Putnam Stewart Shapiro, Tom Ricketts, Sander Verhaegh, and Alan Weir; for Chapter 9, Kate Abramson, David Chalmers, J. Dmitri Gallow, Jim Joyce, Mark Kaplan, Kirk Ludwig, Hilary Putnam, Joshua Schechter, and an anonymous reader for the *Journal of Philosophy*;

for Chapter 10, Adrian Cussins, David Finkelstein, Scott Kimbrough, Art Melnick, Tom Meyer, Hilary Putnam, David Shwayder, Tadeuzs Szubka, Charles Travis, and Steve Wagner; for Chapter 11, Derrick Darby, David Finkelstein, Paul Horwich, Peter Hylton, Michael Kremer, Richard Kraut, Cristina Lafont, Tom McCarthy, Axel Mueller, Hilary Putnam, and Charles Travis. Warm thanks to all.

The Introduction, Chapters 2 and 6, and the Afterword are published in this volume for the first time. Chapters 1, 3–5, and 7–11 were first published elsewhere, as follows:

Chapter 1, "Carnap's Logical Syntax," in Richard Gaskin (ed.), *Grammar in Early Twentieth-Century Philosophy* (London: Routledge, 2001), pp. 218–237.

Chapter 3, "Carnap and Quine on Truth by Convention," *Mind* Vol. 120, No. 478 (April 2011): pp. 193–237.

Chapter 4, "Quine's Naturalistic Explication of Carnap's Logic of Science," in Gilbert Harman and Ernie Lepore (eds.), *A Companion to W. V. O. Quine* (Oxford: Wiley-Blackwell, 2014), 465–482.

Chapter 5, "Quine Gets the Last Word," *Journal of Philosophy*, Vol. 108, No. 11 (November 2011): 617–632.

Chapter 7, "Can Logical Truth Be Defined in Purely Extensional Terms?" *British Journal for the History of Philosophy*, Vol. 22, Issue 2 (2014): 343–367.

Chapter 8, "Reading Quine's Claim That No Statement Is Immune to Revision," in Janssen-Lauret and Kemp (eds.), *Quine and His Place in History* (Houndmills, UK: Palgrave Macmillan Press, 2015), pp. 123–407.

Chapter 9, "Conditionalization and Conceptual Change: Chalmers in Defense of a Dogma," *Journal of Philosophy*, Vol. 111, No. 12, December 2014: 689–703.

Chapter 10, "Truth and Transtheoretical Terms," in James Conant and Urszula Zeglen (eds.), *Hilary Putnam: Pragmatism and Realism* (London: Routledge, 2002), pp. 167–185.

Chapter 11, "Putnam and the Contextually A Priori," in Lewis E. Hahn and Randall E. Auxier (eds.), *The Philosophy of Hilary Putnam* (La Salle, IL: Open Court, 2015), pp. 389–407.

I thank the respective journals, presses, and editors for granting me permission to reprint these articles.

I also thank Hilary Gaskin for her interest in this project and her efficient oversight of the review and contracting for the book; Andrew Smith for his help in preparing electronic typescript files from the published versions of the essays; and Susan Thornton for her careful copyediting.

My greatest debt is to my wife, Martha Lhamon, to whom this book is dedicated, for her unfailing patience, love, and support during the many years that I worked on these essays.

# Introduction

These essays present results of my efforts to understand and learn from Rudolf Carnap's, W. V. Quine's, and Hilary Putnam's writings about fundamental methodological questions, including such questions as whether rational inquiry is or should be governed by rules of language, whether some of our statements are analytic, whether there is a useful or explanatory notion of truth by convention, and whether there are terms whose references are the same despite fundamental differences in the beliefs we and other speakers use the terms to express. The essays present new interpretations of Carnap's, Quine's, and Putnam's answers to such questions; critically evaluate these authors' views in light of the interpretations; and develop new minimalist applications of central philosophical principles suggested by the interpretations.

The essays in Chapter 2 and Chapter 6 were written very recently and are published here for the first time. The other essays, which were published previously, are reprinted here without substantive revisions. In these essays I converted notes and references to a uniform style, changed the wording here and there, and removed minor errors. In the few places where I made changes that modify the content of a claim or argument I added footnotes to explain the changes. I did not reduce repetitions. The focus of each essay differs from that of the others, so the recurring topics are approached differently each time, thereby contributing, I believe, to a deeper understanding.

Taken together, and considered from a standpoint that abstracts from many details, the essays suggest that Carnap, Quine, and Putnam each accept that

> In our pursuit of truth, we can do no better than to start in the middle, relying on already established beliefs and inferences and applying our best methods for reevaluating particular beliefs and inferences and arriving at new ones.

No part of our supposed knowledge, no matter how clear it seems to us or how firmly we now hold it, is unrevisable or guaranteed to be true.

Insofar as traditional philosophical conceptions of reason, justification, and apriority conflict with the first two principles, they should be abandoned. In particular, the traditional philosophical method of conceptual analysis should be abandoned in favor of the method of explication, whereby a term we find useful in some ways, but problematic in others, is replaced by another term that serves the useful purposes of the old term but does not have its problems.

A central task of philosophy is to clarify and facilitate our rational inquiries by replacing terms and theories that we find useful in some ways, but problematic in others, with new terms and theories that are as clear and unproblematic to us as the terms and methods of our best scientific theories.

One characteristic that unifies the essays is that they each take steps toward developing new applications of these four schematic principles. Another is that the steps they take include presenting new interpretations of Carnap's, Quine's, or Putnam's views and developing correspondingly new ways of extending or modifying these views.

The new applications are inspired by Carnap's revolutionary recommendation that we reject first philosophy, according to which philosophy should be based on insights that are "higher" and "deeper" than even the best results of scientific inquiry, and replace it with a positive program for clarifying and facilitating rational inquiry in scientific terms. The new applications differ significantly from Carnap's, however, in several closely connected ways. First, and most important, they are inspired by Quine's and Putnam's central insight that we can reject first philosophy without relying on Carnapian linguistic frameworks, by taking scientific method, loosely characterized, yet concretely applied in our best scientific judgments, as our ultimate arbiter of truth, and drawing on our best current judgments, vocabulary, and methods to clarify and facilitate our inquiries. Second, the new applications incorporate and extend several distinctively Quinean views, including that

We can use our sentences to make assertions, and to support, rescind, or revise our assertions, without implying or presupposing that the meanings of our sentences are fixed by semantical rules.

The notions of satisfaction and truth, as explained in Tarski's way for particular languages, are better suited to serious philosophical work, such as defining logical truth, than any of our commonsense or intuitive philosophical notions of meaning.

Third, the new applications extend these Quinean views in ways that are often assumed to be incompatible with them, by developing and incorporating minimalist extensional explications of some distinctively Putnamian positions, including that

> Our identifications of transtheoretical terms – i.e., terms whose references are the same despite fundamental differences in the beliefs we and other speakers use the terms to express – are integral to our actual practices of agreeing, disagreeing, adjudicating disputes, and revising our beliefs.

> An account of our rational inquiries is adequate only if it fits with these practices.

> The difference between statements that may be overthrown by the outcomes of experiments that we can describe and perform without creating radically new theoretical structures, and statements that may be overthrown only by conceiving of radically new theoretical structures, is of logical and methodological significance, and not just of psychological interest.

A central theme of the essays is that one way to learn from standard criticisms of Carnap's, Quine's, or Putnam's views is to try to reformulate and evaluate the criticisms in terms that the philosophers who are the targets of the criticisms can accept. In some cases, several of which I discuss in the essays, such efforts lead nowhere, or the reformulated criticisms fall flat, revealing that the criticisms are less powerful than they may at first appear. In other cases, several of which I also discuss in the essays, a reformulation of a criticism reveals an overlooked theoretical option, or a new one. One of the overlooked theoretical options revealed and developed in the essays, especially the essay in Chapter 3, is Quine's view that some statements of the most abstract parts of science are true by convention in a thin explanatory sense that does not guarantee truth. Another of the overlooked theoretical options, the focus of Chapter 5, is Quine's resolutely minimalist account of language use, according to which our use of language to make assertions is both pragmatically indispensible to scientific theorizing and independent of particular scientific theories of the semantic relations between words and things.

One of the new theoretical options revealed and developed in the essays, especially the essay in Chapter 7, is to introduce regimented languages by stipulating Tarski-style truth theories for them, thereby licensing us to regard their terms as (extensionally) univocal, while not regarding the stipulations as analytic. Another, presented in Chapter 9, is to combine our practices of identifying transtheoretical terms with revisable Bayesian confirmational commitments. A third, the focus of Chapter 10, is to combine our practical identifications of transtheoretical terms with a minimalist account of truth and reference, thereby fitting our account of truth and reference with our actual practices of identifying agreements and disagreements, adjudicating disputes, and revising our beliefs.

I have grouped and ordered the essays to highlight their development of central themes. Here is a brief overview.

## Carnap

Carnap recommends that we abandon first philosophy and begin inquiry in the midst of our already existing sciences, refining and clarifying them from within. The motivating idea behind his logic of science is that all disputes are resolvable by a series of uncontroversially correct applications of shared rules for inquiry. Chapter 1, "Carnap's Logical Syntax," argues that the only way to challenge, not simply reject, Carnap's logic of science is to question the pragmatic appeal of his movitating idea. The essay argues that this idea leads Carnap to be skeptical of all of our unregimented identifications of transtheoretical terms, even those we rely on in our most exact sciences, and concludes that to reject these identifications is too high a price to pay for the dubious benefit of ruling out the very possibility of engaging in unresolvable disputes. (Chapter 10, "Truth and Transtheoretical Terms," of Part IV, returns to this theme.)

Carnap's classic essay "Empiricism, Semantics, and Ontology" (Carnap 1950, henceforth ESO) presents one instructive way of identifying and eschewing traditional philosophical questions, with a special focus on questions about existence. Chapter 2, "Carnap on Ontology," argues that all of the general existence statements Carnap aims to clarify and defend in ESO – not only those about abstract objects, such as "There are numbers," but also those about concrete objects, such as "There are physical objects" – are, when viewed in the way Carnap recommends in ESO, analytic (that is, settled solely by the rules of the languages in which they are expressed) and trivially so (that is, derivable from the rules in a few simple steps). Chapter 2 also explains why these claims are central to Carnap's method of identifying and eschewing traditional philosophical questions about existence.

## Carnap and Quine

Chapter 3, "Carnap and Quine on Truth by Convention," establishes, among other things, that Quine's debate with Carnap over the analytic–synthetic distinction is rooted in differences in how Quine and Carnap understand Carnap's recommendation that we abandon first philosophy and begin inquiry in the midst of our already-existing sciences, refining and clarifying them from within. Quine's naturalistic interpretation of this new way of doing philosophy leads him to reject Carnap's analytic–synthetic distinction and to posit truth by convention as a transient yet integral part of our current best theory. Chapter 3 thereby shows how Quine appropriates and yet also transforms Carnap's recommendation that we reject the traditional philosophical assumption that there is a legitimate methodological, epistemological, or metaphysical standpoint that is higher or deeper than the standpoint of our best science.

Chapter 4, "Quine's Naturalistic Explication of Carnap's Logic of Science," further develops this interpretation of Quine by arguing that to understand Quine's epistemology, one must see how it incorporates, yet also transforms, Carnap's principled rejection of the traditional empiricist idea that our best theories of nature are justified by, or based on, our sensory evidence, and are for that reason likely to be true. Both Carnap and Quine regard this traditional empiricist idea as rooted in traditional epistemology, which they resolutely reject. Chapter 4 argues that contrary to standard interpretations of Quine, the goal of Quine's naturalistic account of the relationship between theory and evidence is not to show that our best theories of nature are justified by our sensory evidence, but to show that we can describe science from within science in a way that mirrors and thereby clarifies Quine's doctrinal principle (itself an explication of Carnap's methodological rejection of metaphysics, the topic of Chapter 2) that we can only judge truth from the standpoint of our best current theory.

## Quine

Chapter 5, "Quine Gets the Last Word," addresses two fundamental kinds of criticisms of Quine's effort to work out a naturalistic, Carnapian rejection of traditional metaphysics, without relying on the analytic–synthetic distinction. The first criticism is that Quine's naturalistic epistemology is descriptive, not normative, and so fails to provide an account of epistemic justification. The second is that Quine's naturalistic account of meaning leads to the absurd conclusion that our sentences are meaningless, so we cannot use them to make assertions. Chapter 5 argues that the entrenched criticisms ultimately rest on a failure to grasp Quine's resolutely minimalist characterization of language use. According to Quine, Chapter 5 argues, insofar as we are using, and not also mentioning, a given string of nonsemantic words, no particular relation between those words and things is asserted, presupposed, or implied. Thus understood, our use of language to make assertions is not undermined by Quine's naturalistic account of meaning. Chapter 5 also argues that if we focus on what Quine tells us about epistemology from the standpoint of our own evolving theory, we find a minimalist, context-dependent sort of normativity that goes hand in hand with Quine's minimalist characterization of language use.

The criticisms that Chapter 5 refutes are based on assumptions about language use that Quine explicitly rejects. Chapter 6, "Reading Quine's Claim That Definitional Abbreviations Create Synonymies," and Chapter 7, "Can Logical Truth Be Defined in Purely Extensional Terms?" investigate two apparently powerful criticisms of Quine's naturalism that question the internal coherence of his naturalistic transformation of Carnap's radical rejection of traditional philosophical methods of inquiry. In "Two Dogmas of Empiricism" Quine claims that definitional abbreviations create transparent synonymies and that all other species of synonymy are less intelligible (Quine 1953a, p. 26). Against this, Grice and Strawson assert that "the notion of synonymy by explicit convention would be unintelligible if the notion of synonymy by usage were not presupposed" (Grice and Strawson 1956, pp. 152–153). This superficially plausible claim has been repeated many times in the literature on Quine and is now among the standard reasons that philosophers give for rejecting Quine's arguments in "Two Dogmas." Chapter 6 demonstrates, however, that Quine's claim about abbreviations does not presuppose a general notion of synonymy by usage, as Grice and Strawson claim, but is trivially derivable from his starting assumption that first-order logical truths, defined purely extensionally, are analytic, together with a commonsense observation about conventional abbreviations. Chapter 6 concludes that Quine's claim that definitional abbreviations create synonymies is not in conflict with his skepticism about the general notion of synonymy.

The argument in Chapter 6 takes for granted that Quine successfully defines first-order logical truth extensionally. Chapter 7 investigates a little-known but apparently powerful internal criticism of Quine's efforts to define logical truth extensionally. Strawson argues that one cannot say what it is for a particular use of a sentence to exemplify a logical form without appealing to intensional notions, and hence that Quine's efforts to define logical truth in purely extensional terms cannot succeed. Chapter 7 shows that Quine's reply to this criticism of Strawson's is confused and develops a better reply using resources that Quine can accept. The better reply is that we can introduce regimented languages by stipulating Tarski-style truth theories for them. From the stipulations we can then easily infer that the terms of the regimented languages are (extensionally) uni-vocal. Chapter 7 thereby answers Strawson's internal challenge, and, as a consequence, also defends the argument of Chapter 6 from the criticism that Quine's easy way with definitional synonymies presupposes a prob-lematic conception of logical truth.

Chapter 8, "Reading Quine's Claim That No Statement Is Immune to Revision," addresses a very different but equally fundamental misun-derstanding of Quine's appropriation of Carnap's rejection of traditional philosophy: most critics and defenders of Quine's arguments in "Two Dogmas of Empiricism" read his claim that "no statement is immune to revision" as the claim that for *every statement S that we now accept there is a possible rational change in beliefs that would lead one to reject S*. Against this standard reading, Chapter 8 argues that Quine's claim is that

(P) No statement we now accept is guaranteed to be part of every scientific theory that we will later come to accept.

The key to the alternative interpretation presented in Chapter 8 is to see that in paragraph 1 of section 6 of "Two Dogmas of Empiricism" Quine sketches a bold new naturalistic explication of the traditional notion of empirical confirmation, and that his aim in paragraph 2 is to show that the explication of confirmation he sketches in the previous paragraph is of no help in characterizing a boundary between analytic and synthetic state-ments. Chapter 8 argues that contrary to the standard view, Quine agrees with defenders of the analytic–synthetic distinction that there are some sentences we cannot now make sense of rejecting without changing their meanings, and argues that this fact is not relevant to the question whether the statements expressed by those sentences are analytic in any philosoph-ically substantive epistemological sense. Chapter 8 therefore shows that one main source of the currently entrenched resistance to Quine's way of developing Carnap's rejection of traditional metaphysics and epistemol-ogy is based on a misunderstanding of Quine's position.

**Quine and Putnam**

Chapter 9, "Conditionalization and Conceptual Change: Chalmers in Defense of a Dogma," considers a recent attempt to refute what many philosophers take to be the central challenge to analyticity – the claim that any sentence can be revised without changing its meaning. The previous essay, Chapter 8, argues that Quine is not committed to this claim, but he is also not committed to rejecting it, and his commitment to fallibilism implies that he cannot rule it out. The claim is central to Putnam's development of Quine's challenges to analyticity, however, and so Chapter 9 is relevant to evaluating both Quine's and Putnam's views on analyticity. David Chalmers argues that Bayesian conditionalization is a constraint on conceptual constancy, and that this constraint, together with standard Bayesian considerations about evidence and updating, is incompatible with the Quinean claim that every belief is rationally revisable. Chalmers's argument presupposes that the sort of conceptual constancy that is relevant to Bayesian conditionalization is the same as the sort of conceptual constancy that is relevant to the claim that every belief is rationally revisable. To challenge this presupposition Chapter 9 explicates a kind of "conceptual role" constancy that a rational subject could take to be necessary and sufficient for a rule of Bayesian conditionalization to govern her belief updating, and shows that a rational subject may simultaneously commit herself to updating her beliefs in accord with such a rule and accept the claim that every belief is rationally revisable.

Chapter 10, "Truth and Transtheoretical Terms," discusses Putnam's central objection to Quine's deflationary view of truth and reference – that it apparently leads to the absurd conclusion that two speakers cannot genuinely agree or disagree with each other. What Putnam's central argument against Quine shows, however, if we reconstruct it in the way Chapter 10 suggests, is that an account of truth and reference is satisfactory only if it fits with and makes sense of our actual practices of agreeing and disagreeing, adjudicating disputes, and revising our beliefs. Chapter 10 sketches a new minimalist account of truth and reference that meets this condition.

## Putnam

Chapter 11, "Putnam and the Contextually Apriori," explores a central question for a Quinean methodology of inquiry: what is the methodological or epistemological significance of our current inability to specify how a statement may actually be false? Clearly, our current inability to specify how a statement may actually be false does not guarantee that we will never be able to do so. Nevertheless, according to Putnam, when we cannot specify how a statement may actually be false it is epistemically reasonable for us to accept the statement without evidence and hold it immune from disconfirmation. Against this, one might be inclined to reason as follows:

> It is epistemically reasonable for a person to accept a particular statement only if she has epistemic grounds for accepting it. But a person's inability

to specify a way in which a statement may actually be false gives her no epistemic grounds for accepting it. Therefore, if the epistemic role of the statement for her is exhausted by her inability to specify a way in which the statement may actually be false, it is not epistemically reasonable for her to accept it.

Chapter 11 argues, on the contrary, that

> If a person has not irresponsibly ignored clues or hints about how to specify a way in which a statement S that she accepts may be false, and she cannot make sense of doubting S, then it is epistemically reasonable for her to accept S.

Chapter 11 also argues that Putnam's efforts to explain why this statement is true are ultimately unsuccessful, and that the failure of Putnam's efforts reveals that the statement is not best viewed as an empirical generalization about the psychology of inquirers, but as a central methodological principle the endorsement of which can help inquirers to stay focused on truth and thereby resist the siren call of first philosophy.

# PART I

## *Carnap*

CHAPTER I

# Carnap's Logical Syntax

Carnap's logical syntax is a systematic expression of his attitude toward different types of discourse.[1] He avoids the vocabulary and methods of traditional metaphysics and endorses the vocabulary and methods of science.[2] His *motivating attitude* is that it is "sterile and useless" to say two investigators agree or disagree unless we see them as sharing criteria for evaluating their assertions.[3]

This attitude led Carnap to seek a method for specifying intersubjective criteria for evaluating logical and mathematical statements. He rejected Wittgenstein's view that all attempts to specify the meanings of logical statements and terms end in nonsense (Carnap 1937, pp. 282–284; Carnap 1935, p. 37). At the same time, he was inspired by (what he regarded as) Wittgenstein's idea that logical truths are "based solely on their logical structure and on the meaning of the terms" (Carnap 1963a, p. 25).[4] Carnap found a way to articulate this idea in January 1931, six months after he learned of Kurt Gödel's incompleteness theorems and the

---

[1] My interpretation of Carnap is deeply indebted to Ricketts 1982, Ricketts 1994, Ricketts 1996, and Goldfarb and Ricketts 1992.

[2] Carnap reports that "the anti-metaphysical attitude showed itself chiefly in the choice of the language used in the discussion. We [members of the Vienna Circle] tried to avoid the terms of traditional philosophy and to use instead those of logic, mathematics, and empirical science, or of that part of the ordinary language which, though more vague, still is in principle translatable into a scientific language" (Carnap 1963a, p. 21). In Carnap 1934, Carnap stresses that "there is nothing else, nothing 'higher' to be said about things than what science says about them" (Carnap 1934, p. 47).

[3] "Most of the controversies in traditional metaphysics appeared to me sterile and useless. When I compared this kind of argumentation with investigations and discussions in empirical science or in the logical analysis of language, I was often struck by the vagueness of the concepts used and by the inconclusive nature of the arguments. I was depressed by disputations in which the opponents talked at cross purposes; there seemed hardly any chance of mutual understanding, let alone of agreement, because there was not even a common criterion for deciding the controversy" (Carnap 1963a, pp. 44f.).

[4] Carnap thought that, properly explicated, Wittgenstein's insight enables us "to combine the basic tenet of empiricism with a satisfactory explanation of the nature of logic and mathematics" (Carnap 1963a, p. 47).

ingenious method of arithmetization that Gödel devised to prove them
(Carnap 1963a, p. 53).

In *The Logical Syntax of Language* (Carnap 1937), Carnap's first detailed
presentation of his new vision of logical syntax, he shows how to construct syntactic definitions of 'logical consequence' for artificial languages.
He proposes that we *replace* the vague sentence "the truth of logical statements is based solely on their logical structure and on the meaning of the
terms" with a syntactical definition: a sentence *s* of a (noncontradictory)
language *L* is logically true (analytic) if and only if *s* is a (special sort of)
syntactical consequence in L of the empty set of sentences of *L*. Using
Gödel's method of arithmetization, Carnap shows how his proposed syntactical definition can be formulated without using any vague vocabulary.

My central goal in this chapter is to explain in more detail how Carnap
proposes to avoid vague discourse by endorsing only the vocabulary and
methods of science. I will explain how Carnap could plausibly see himself
as the practitioner of a new *scientific* discipline – *Wissenschaftslogik*, or the
logic of science. In 1934 he boldly announced that

> our own discipline, logic or the logic of science, is in the process of cutting
> itself loose from philosophy and becoming a properly scientific field, where
> all work is done according to strict scientific methods and not by means of
> 'higher' or 'deeper' insights. (Carnap 1934, p. 46)

If there is anything resembling philosophy in Carnap's new discipline, it
is his motivating attitude, which underlies his rejection of discourse that
he finds unclear. The only way to challenge, not simply reject, Carnap's
motivating attitude is to question its pragmatic appeal. In the last two sections of the chapter, I will suggest that Carnap asks us to pay too high a
price for the dubious benefit of ruling out the very *possibility* of engaging
in unresolvable disputes.

## 1.1   Formation and Transformation Rules

The primary task of the logic of science is to construct and describe *language systems*. Each language system *S* is characterized by its *formation
rules*, which specify the sentences of *S*, and its *transformation rules*, which
together settle, for every sentence *s* of *S* and every set *R* of sentences of *S*,
whether or not *s* is a *consequence* of *R* in *S*.

Carnap's conception of formation rules is straightforward. It is his
conception of transformation rules that is distinctive of his logical syntax. Given a complete set of formation rules for a language system *S* that

contains the symbol '→', we may stipulate that *modus ponens* is among the transformation rules of *S*:

> If *s* and *r* (hence also ⌜*s* → *r*⌝) are sentences of *S*, then *r* is a consequence of {*s*, ⌜*s* → *r*⌝} in *S*.

According to Carnap, the transformation rules for *S* may also include a list of primitive sentences, or axioms, such as ⌜*s* → (¬*s* → *r*)⌝, for all sentences *s* and *r* in *S*. To put a sentence *s* of *S* on the list of primitive sentences of *S* is to stipulate that *s* is a consequence in *S* of the empty set of sentences of *S*. Taken together, the transformation rules for *S* settle what Carnap calls the *consequence* relation for *S*.

Transformation rules for a language system *S* may include axioms of arithmetic or set theory; so the consequence relation for *S*, as Carnap defines it, may be considerably stronger than the consequence relation for a typical first-order quantificational language, for instance. Moreover, Carnap's consequence relation for a system *S* cannot in general be the same as the *derivability* relation for *S*, which is settled by *finite* transformation rules ("primitive sentences and rules of inference each of which refers to a finite number of premises," Carnap 1939, p. 165). Gödel's first incompleteness theorem shows that no set of *finite* transformation rules can capture our pretheoretical notion of logical consequence for language systems that contain arithmetic. Carnap knew this and proposed that we allow *transfinite* rules of transformation, which are defined for an infinite number of premises (Carnap 1937, §§14, pp. 34, 43–45). According to the transfinite rule DC2 of Carnap's Language I, for instance, a sentence *s* of Language I that contains some free variable *v* is a logical consequence of the infinite class of sentences of Language I that is obtained by substituting names of positive integers for *v* in *s* (Carnap 1937, p. 38; Sarkar 1992, p. 197). Carnap calls such transfinite rules *syntactical* because they can be specified formally, even though the question whether they are correctly applied in a given case is not decidable (Carnap 1937, p. 100).

The formation and transformation rules for a language system can be specified by Gödel's method of arithmetization. The method exploits the fact that each number has a unique factorization into primes. If each symbol in the language is assigned a distinct number, the sentences of the language can be put into correspondence with numbers in such a way that the sentence represented by a given number can be recovered by factoring that number into primes. The formation rules for a language system can be specified by an arithmetical definition of the set of all and only those numbers that correspond in this way with sentences of the

system. Similarly, the transformation rules for the system can be specified by an arithmetical definition of the set of all and only those numbers
that correspond with series of numbers that correspond with derivations
of sentences in the system. In this way the formation and transformation
rules of the system can be specified in purely arithmetical terms (Carnap
1937, p. 57).

By arithmetizing the formation and transformation rules for language
systems, Carnap satisfies his requirement that his proposed syntactical
specifications of the logic of science be expressed in exact scientific terms.
For instance, suppose we want to assert that a particular sentence $s$ is not
provable in language system $S$. Roughly speaking, '$s$ is not provable in $S$'
amounts to 'No proof of $s$ is *possible* in $S$'. To allay any doubts about the
clarity of this notion of possibility, we can arithmetize the syntactical rules
of $S$, and *replace* 'No proof of $s$ is possible in $S$' with a sentence of the form
'There exists no number that corresponds with a proof of $s$ in $S$' (Carnap
1937, p. 57).

## 1.2   Some Central Terms of Logical Syntax

Starting with definitions of transformation rules, which settle the *consequence* relation for a language system $S$, Carnap constructs a series of definitions that he proposes that we use in place of our intuitive ideas about
the logical grounds for affirming our sentences. Like formation and transformation rules, the new definitions may be arithmetized, and therefore
have none of the vague intuitive significance of the phrases they replace.
The series begins as follows:[5]

> Sentence $s$ of language system $S$ is valid if and only if $s$ is a consequence in
> $S$ of the empty set of sentences of $S$; and contravalid if and only if every
> sentence of $S$ is a consequence of $s$ in $S$.

For any sentence $s$ of a *noncontradictory* language system $S$, Carnap proposes that we use '$s$ is valid in $S$' in place of '$s$ is true in virtue of the rules
of $S$', and '$s$ is contravalid in $S$' in place of '$s$ is false in virtue of the rules of
$S$'.[6] Then we can replace '$s$ is true or false in virtue of the rules of $S$' with '$s$
is determinate', defined as follows:

---

[5] I follow Carnap's exposition of these definitions in Carnap 1935a, ch. II.
[6] Carnap does not accept this replacement for sentences of contradictory language systems. See §1.6
and n. 14.

Sentence *s* of language system *S* is *determinate* if and only if either *s* is *valid* or *s* is *contravalid*.

It follows from this definition that *s* is *not determinate* (*indeterminate*) if and only if *s* is neither *valid* nor *contravalid*. The next few definitions depend on Carnap's distinction between two types of transformation rules: L-rules and P-rules. All language systems contain L-rules; some language systems, but not all, contain P-rules. The L-rules of a language system *S* settle which sentences are logically true or false in *S*. Together with the L-rules of *S*, the P-rules of *S* settle which sentences are determinate but neither logically true nor logically false. Any nonlogical sentence can be defined as a primitive P-rule. Carnap proposes that we include sentences that express laws of nature among the P-rules of language systems we use for physics, so that the sentences are *valid* – "true in virtue of the rules" – and hence *determinate* in those systems (Carnap 1935a, pp. 51–54). Assume that the distinction between the L-rules and P-rules of each language system *S* is drawn *within S* by the definitions of the L-rules and P-rules of *S*.[7] Given this distinction, Carnap constructs the following series of definitions:

A sentence *r* is an *L-consequence* of a sentence *q* in a system *S* if and only if the L-rules of *S* together settle that *r* is a consequence of *q*.

Sentence *s* of language system *S* is *analytic* if and only if *s* is an L-consequence in *S* of the empty set of sentences of *S*; and *contradictory* if and only if every sentence of *S* is an L-consequence of *s* in *S*.

Sentence *s* of language system *S* is *L-determinate* if and only if either *s* is *analytic* or *s* is *contradictory*; and *synthetic* if and only if *s* is not *L-determinate*.

Carnap stresses that for each language system we can use Gödel's method to define all of these syntactical terms arithmetically.

## 1.3 Limits on Defining the Logical Syntax of *S* within *S*

How much of the logical syntax of a language system *S* can be defined within *S*? Can 'analytic (in S)' be defined in *S*? Carnap proves that the answer to this second question is "No." Using techniques he learned from

---

[7] In §51 of Carnap 1937, Carnap offers a general syntactical criterion for partitioning the transformation rules of any language system *s* into the set of L-rules and the set of P-rules of S. He abandoned the criterion after he embraced Tarski's method for defining truth. See Carnap 1942, §39, p. 247f. According to Michael Friedman (Friedman 1988 and 1999a), Carnap needs a general syntactical criterion for distinguishing between the L-rules and P-rules of a language system. I do not have the space here to discuss this central issue.

Gödel, Carnap sketches a syntactical version of the liar paradox to show that *if S is consistent, then 'analytic (in S)' is indefinable in S* (Theorem 60C.I, Carnap 1937, p. 219). His proof depends on two assumptions: (1) for every arithmetical sentence *s* of a *noncontradictory*[8] language system *S*, '*s* is true (false) in *S*' can be translated as '*s* is analytic (contradictory) in *S*', and (2) "it is possible to construct, for any and every syntactical property formulable in *S*, a sentence of *S* ... that ... attributes this property – rightly or wrongly – to itself" (Carnap 1937, p. 217). Assumption (2) is Carnap's fixed-point theorem; he proved it by generalizing from Gödel's method of constructing an arithmetical sentence *s* of a system *S* that is true if and only if *s* is not provable in *S* (Carnap 1937, pp. 129–131; Sarkar 1992, p. 205).

Here is a sketch of Carnap's proof. Assume that *S* is noncontradictory (consistent) and contains arithmetic. Suppose (toward contradiction) that 'analytic in *S*' is defined in a syntax formulated in *S*. Then 'not analytic in *S*' can also be defined in *S*. By assumption (2), it is therefore possible to construct an arithmetical sentence *s* of *S* such that, relative to the arithmetization, *s* "says" that *s* is not analytic in *S*. If *s* is not analytic in *S*, then since *s* is a sentence of arithmetic, *s* is contradictory in *S*; hence by assumption (1), *s* is false in *S*. Since *s* "says" that *s* is not analytic in *S*, and *s* is false in *S*, we may conclude that it is not the case that *s* is not analytic in *S*; hence *s* is analytic in *S*. But if *s* is analytic in *S*, then by assumption (1), *s* is true in *S*. Since *s* "says" that *s* is not analytic in *S*, and *s* is true in *S*, we may conclude that *s* is not analytic in *S*. In short, *s* is analytic in *S* if and only if *s* is not analytic in *S*. This is a contradiction. Hence '*s* is analytic in *S*' cannot be defined in *S*.[9]

In this respect, '*s* is analytic in *S*' contrasts with '*s* is provable in *S*'. From Gödel's first incompleteness theorem we know that there is no contradiction in the supposition that there is a language system *S* in which '*s* is provable in *S*' can be arithmetically defined. To see why, assume that *s* is noncontradictory (consistent) and contains arithmetic, and suppose that '*s* is provable in *S*' is defined in *S*. By assumption (2), it is then possible to construct an arithmetical sentence *s* of *S* such that, relative to the arithmetization, *s* "says" that *s* is not provable in *S*. Suppose *s* is false in *S*. Then since *s* is a sentence of arithmetic, *s* is contradictory in *S*, by assumption (1). And since *s* "says" that *s* is not provable in *S*, and *s* is false in *S*,

---

[8] This qualification is important in Section 1.6, where I claim that according to Carnap no sentence of a contradictory language system is true.

[9] Carnap's sketch of this proof is slightly different; see Carnap 1937, §60c, p. 218.

it is not the case that *s* is not provable in *S*; hence *s* is provable in *S*. But then *s* is both provable in *S* and contradictory in *S*, contrary to the supposition that *s* is consistent. So *s* must be true in *S*. Since *s* "says" that *s* is not provable in *S*, and *s* is true in *S*, we may conclude that *s* is not provable in *S*. Hence *s* is true in *S* but not provable in *S*. This is not a contradiction. What it shows is that *S* contains at least one true arithmetical sentence that is not provable in *S*.[10]

From the reasoning of this sort (more precisely, from Gödel's proof of his first incompleteness theorem), Carnap concludes that "everything mathematical can be formalized, but mathematics cannot be exhausted by one system; it requires an infinite series of ever richer languages" (Carnap 1937, p. 222). And from reasoning of the sort sketched two paragraphs above, Carnap concludes that "if the syntax of a language $S_1$ is to contain the term 'analytic (in $S_1$)' then it must, consequently, be formulated in a language $S_2$ which is richer in modes of expression than $S_1$" (Carnap 1937, p. 219). For technical reasons internal to Carnap's project of making logical syntax scientifically explicit, we must acknowledge that there are infinitely many language systems, no one of which is sufficient to express the logical syntax of *all* language systems.

## 1.4 Carnap's Principle of Tolerance

One might be tempted to ask, "Which language system is correct?" Carnap rejects this question. He endorses a Principle of Tolerance, according to which

> Everyone is at liberty to build up his own logic, i.e. his own form of language, as he wishes. All that is required of him is that, if he wishes to discuss it, he must state his method clearly, and give syntactical rules instead of philosophical arguments. (Carnap 1937, p. 52)

Given the Principle of Tolerance, Carnap can express his *motivating attitude* – that it is "sterile and useless" to say two investigators agree or disagree unless we see them as sharing criteria for evaluating their assertions – by recommending that we use *only* those languages whose transformation rules can be specified syntactically.

He sees the Principle of Tolerance itself as a characteristically scientific attitude, "shared by the majority of mathematicians" (Carnap 1937, p. 52).

---

[10] For Carnap's account of the disanalogy between '*s* is analytic in *S*' and '*s* is provable in *S*' see Carnap 1937, §60c, pp. 217f.

This is not just wishful thinking; in actual mathematical practice one can find many illustrations of Carnap's Principle of Tolerance. For instance, in set theory there are conflicting but equally acceptable ways of defining 'ordered pair'. And number theorists sometimes find it useful to define 'integer' in such a way that some imaginary numbers, such as the square root of negative 1, are 'integers'. Commenting on a proposed definition of 'integer' for quadratic fields, the mathematician Harold Stark writes

> A definition should not be condemned just because it is unnatural. One should not pass judgment on it until one sees if it leads to new results and clarifies old ones. (Stark 1995, p. 265)

This passage expresses a pragmatic attitude toward mathematical definitions, according to which an "unnatural" definition may be acceptable if "it leads to new results and clarifies old ones."

Since the definitions of logical syntax can be formulated arithmetically, the mathematicians' tolerant attitude toward arithmetical definitions applies to the definitions of logical syntax. *By means of arithmetization, the attitude of tolerance that is implicit in mathematical practice can be seen to encompass the definitions of logical syntax.*[11] In this scientific spirit of tolerance, Carnap proposes that we replace the old question "How do we know logical and mathematical truths?" with two new questions: "Which language systems should we use?" and "Which sentences of the language system that we have decided to use are L-valid?" The first question is *pragmatic*; there is no correct or incorrect answer to it. The second question is *mathematical*; we address and resolve it by investigating the consequences of the arithmetical definitions of logical syntax. Carnap sometimes says that the L-valid and L-contradictory sentences of a language system "have no factual content." From a methodological perspective, this means that the L-valid and L-contradictory sentences are "mere calculational devices" that are constructed so that they can be easily applied to the L-indeterminate sentences (Carnap 1935b, p. 126). Among the L-indeterminate sentences, in contrast, are sentences that express particular "matters of fact," in the sense that they can be confirmed or disconfirmed by empirical observations expressed by using what Carnap calls *protocol sentences*.

---

[11] Viewed in isolation, many definitions in logical syntax are mathematically uninteresting. Put together in the right way, however, definitions in logical syntax enable us to prove interesting new mathematical results, such as Gödel's incompleteness theorems and Carnap's fixed-point theorem. Such definitions therefore lead to new mathematical results and clarify old ones. I am grateful to Joan Weiner and Michael Friedman for helpful discussions of this point.

To illustrate these points, Carnap constructs two languages: Language I, which contains only decidable predicates of the natural numbers, and Language II, which includes Language I but also contains undecidable predicates of the natural numbers. Carnap shows how to construct mathematical definitions of 'L-consequence in Language I' and 'L-consequence in Language II'. Both language systems contain analytic mathematical sentences. But they also "afford the possibility of constructing empirical sentences concerning any domain of objects" (Carnap 1937, p. 12). For instance, to Language II we can add some terms and predicates of physics and extend the formation rules to construct sentences of a new language system, Language $II_P$. Some of the sentences of Language $II_P$ will be indeterminate. Carnap recommends that the terms and predicates of physics that we add to Language $II_P$ be chosen so that any two investigators who use Language $II_P$ agree on how to evaluate its indeterminate sentences, including its protocol sentences, "by means of which the results of observation are expressed" (Carnap 1937, p. 317).

## 1.5 Carnap's AntiMetaphysical Attitude

We can now say more explicitly how Carnap eschews the vocabulary and methods of metaphysics. If $s$ is a sentence of metaphysics, then not all investigators agree on how to evaluate $s$. Carnap proposes that we avoid the vocabulary and methods of metaphysics by using only those language systems $S$ such that for every sentence $s$ of $S$, all investigators agree on how to evaluate $s$.

One might think that there are "contentful" metaphysical claims that cannot be expressed by sentences of any language system of the sort that Carnap endorses. To clarify the question of whether there are such claims, Carnap proposes the following definitions:

> The cognitive content of sentence s in language system $S$ is the class of the non-valid sentences that are consequences of s in $S$. (Carnap 1937, p. 175)
>
> A sentence s in S has (a) the *null content* if and only if its cognitive content is the empty set, and (b) the *total content* if and only if its cognitive content is the set of *all* the sentences of S. (Carnap 1937, p. 176)

It follows from these definitions that every valid sentence of $S$ has the null content, and every contravalid sentence of $S$ has the total content. The vague idea that "$s$ is contentful" corresponds with the syntactical claim that "$s$ has a nonempty cognitive content that does not include all the sentences of the language system $S$ to which $s$ belongs" – "the cognitive content of $s$ is factual," for short.

To answer the question whether the cognitive content of $s$ is factual, we must correlate it with a sentence in a language system that we can specify (Carnap 1937, p. 8).[12] By definition, any sentence with "cognitive content" can be correlated with (and, relative to the correlation, is expressible by) a "factual" sentence of a precisely specified language system. Given these definitions, the initially tempting thought that there are contentful metaphysical sentences which cannot be expressed by a sentence of one of Carnap's constructed language systems amounts to the *contradictory* thought that there are sentences whose cognitive content is "factual" – sentences that by definition all investigators agree on how to evaluate – that not all investigators agree on how to evaluate.

If philosophy is *replaced* by the logic of science, as Carnap recommends, then there are no contentful "philosophical theses." Such theses, if we can make sense of them at all, may be paraphrased either by determinate sentences of logical syntax, such as "'five' is a number word in language system $S$," or by explicitly formulated *proposals* as to which language systems we should use (Carnap 1937, pp. 286, 299). The former can be evaluated by the methods of logical syntax, whereas the latter – including Carnap's proposal that we use language systems that contain only sentences that all investigators agree on how to evaluate – are neither true nor false, and can be evaluated by an individual only on pragmatic grounds, relative to his or her interest and goals. If one's goals are to make new discoveries, clarify old discoveries, and avoid fruitless disputes, for instance, one may decide to use only those language systems whose rules can be specified syntactically in the way that Carnap recommends.

## 1.6   Acceptance, Consistency, and Truth

Despite Carnap's antimetaphysical rhetoric, many believe he is committed to *conventionalism* about logical and mathematical truth, i.e., to the thesis that our *acceptance* of logical and mathematical sentences somehow

---

[12] If the sentence in question is part of a constructed language system, the correlation we need is just the identity relation. If the sentence in question is part of an historically given language, then the correlations that we need fall under the province of *descriptive syntax*, within which investigators specify language systems suitable for raising *empirical* questions about the cognitive contents of sentences of historically given languages. I discuss Carnap's account of descriptive syntax in Ebbs 1997, ch. 4.

*guarantees* that those sentences are true.[13] In fact, however, Carnap himself proved that there can be no such guarantee. To see this, it helps to see first what is wrong with Kurt Gödel's criticism of Carnap's view that the truth of mathematical sentences is settled by rules of logical syntax.

Gödel's criticism is that we cannot know on *syntactical* grounds alone whether or not a given sentence expresses a truth. A key consideration is that

> [A] rule about the truth of sentences can be called *syntactical* only if it is clear from its formulation, or if it somehow can be known beforehand, that it does not imply the truth or falsity of any "factual" sentence. (Gödel 1995, p. 339)

Only a *consistent* rule will satisfy this requirement, since an inconsistent rule implies *all* sentences, including "factual" ones. But Gödel's second incompleteness theorem shows that the consistency of a system $S$ of arithmetic cannot be proved using only rules and expressions that can be formulated within $S$. Hence the truth of mathematical sentences is not settled solely by the explicit rules of logical syntax.

On Warren Goldfarb's interpretation, this objection presupposes that there is a notion of "fact" that is available independent of particular language systems:

> As Gödel characterizes the positivist view, first there are empirical sentences, which are true or false by virtue of facts in the world; mathematics is then added, by means of conventional syntactical rules. Gödel's argument is that this addition has to be known not to affect the empirical sentences given at the start, and, by his theorem, to ascertain that requires more mathematics. Hence there is a *petitio*. (Goldfarb 1996, p. 226)

Goldfarb replies that

> in this [Carnap's] view, there is no admission of "fact" or "empirical world" that is given prior to linguistic frameworks... The notion of empirical fact is given *by way* of the distinction between what follows from the rules of [a] particular language and what does not, so that different languages establish different domains of fact. In this way, Carnap's view undercuts the very formulation of Gödel's argument. (Goldfarb 1996, p. 227)

If Gödel's argument presupposes a notion of fact that is given prior to the rules of a language system, then Goldfarb's reply is decisive.

---

[13] One influential source of this view is Quine 1963a. For reasons I explain in Chapter 3 of this volume, however, I now think that, despite what many readers believe, Quine did not attribute this view to Carnap.

But there is a related objection to Carnap's approach that does not pre-suppose a notion of fact that is given prior to any language system. This related objection takes for granted, instead, that according to Carnap *our acceptance of a sentence s solely on the basis of the L-rules of a language system S that we take to be consistent guarantees that s is true in S*. The objection is that our *acceptance* of certain syntactical rules, including our acceptance of certain sentences as 'primitive truths' of a given language system, does not by itself determine that those rules or sentences are consistent. A true sentence must be consistent, but Gödel's second incompleteness theorem shows that our acceptance of syntactical rules cannot by itself guarantee their consistency.

The trouble with this objection is that Carnap did not accept the thesis it undermines. He takes for granted that a contradictory sentence is not true,[14] and he states explicitly that our acceptance of syntactical rules cannot by itself guarantee their consistency (Theorem 60C.2, Carnap 1937, p. 219). For instance, after proving the consistency of his Language II, Carnap notes that "since the proof is carried out in a syntax-language which has richer resources than Language II, we are in no wise guaranteed against the appearance of contradictions in this syntax-language, and thus in our proof" (Carnap 1937, p. 129). This means that even if we have proved that language system S is consistent, using a stronger mathematical language whose consistency we take for granted (but have not proved), there is no guarantee that if sentence s is analytic in S, then s is consistent in S, hence no guarantee that if sentence s is analytic in S, then s is true in S.[15]

Carnap's response to the possibility of inconsistency is pragmatic. Suppose we feel confident that a given language system S is both more useful to us and more likely to be contradictory than another. Even if we wish to avoid inconsistency, we may decide that S's usefulness is worth the greater risk of contradiction.[16] Gödel might reply that the transfinite rules

---

[14] Carnap says that "If the truth of the sentence in question follows from the rules of transformation of the language in question, then 'true' can be translated by 'valid' (or, more specifically, by 'analytic'...) and correspondingly, 'false' by 'contravalid' (or 'contradictory'...)" (Carnap 1937, pp. 216f.). I read Carnap's 'if' here as the traditional mathematician's 'if and only if' and assume that the biconditional does not hold for analytic sentences of contradictory language systems.

[15] Since there can be no indeterminate sentences in an inconsistent language system, Carnap is also committed to denying that if we can prove, on the assumption that language system S is consistent, that a given sentence s of S is *not* deducible solely from the L-rules of S, then we are guaranteed that s is indeterminate in S.

[16] This attitude is exemplified by Carnap's willingness to use classical mathematics, even though he acknowledges that intuitionistic mathematics is much less likely to be contradictory (Carnap 1939, p. 193).

that Carnap needs in his "syntactical" definitions of logical consequence for language systems that contain arithmetic are not truly *syntactical*, since (as Carnap concedes) we have no guarantee that we have correctly applied them in any given case. But isn't Carnap entitled to use the word 'syntactical' as he likes, provided that he has made his meaning clear? The key point is that Carnap explicitly acknowledges that our acceptance of a language system in no way guarantees that it is consistent or that any of its sentences is true.[17] Contrary to what many believe, he rejects the confused thesis that our acceptance of logical and mathematical sentences somehow guarantees that those sentences are true.

## 1.7 Rules

The myth that Carnap is committed to conventionalism about logical and mathematical truth goes hand in hand with misunderstandings of his conception of the rules of a language system. Michael Dummett expresses one such misunderstanding when he suggests that Wittgenstein's remarks on rule-following in *Philosophical Investigations*, if correct, would undermine the very idea of rule-governed language systems of the sort that Carnap endorses (Dummett 1993, pp. 446–461). Given Carnap's arithmetization of logical syntax, Dummett is right about this only if Wittgenstein's remarks on rule-following conflict with Carnap's assumptions about arithmetic.

Dummett thinks Wittgenstein espouses a radical conventionalism, according to which every new application of an arithmetical expression embodies a new "decision" about how to apply the expression, and nothing "determines in advance" whether any such decision is correct or incorrect. In contrast, according to Dummett, Carnap and the other logical positivists took for granted that the rules of arithmetic "determine in advance" what counts as the correct application of arithmetical expressions (Dummett 1993, p. 447). As we have seen, however, Carnap is not interested in or committed to any theses that cannot in principle be settled by scientific methods codified by a particular language system. By hypothesis, neither the supposed thesis that every new application of an arithmetical expression embodies a new decision about how to apply the expression nor the supposed thesis that arithmetical rules determine in

---

[17] Note also that the syntactical rules of a *consistent* language system $S$ do not somehow 'make' the analytic sentences of $S$ true. Even if language system $S$ is consistent, the most we can say is that if we *know* that a sentence $s$ is analytic in $S$, then we cannot refuse to affirm $s$ without thereby also refusing to use language system $S$.

advance what counts as the correct application of arithmetical expressions can be evaluated by scientific methods. By Carnap's standards, then, neither supposed thesis has any cognitive content.

Carnap might be willing to say that the rules of a given language system $S$ "determine" that some sentence $s$ is analytic in $S$. But for him this would amount to an *arithmetical* statement that does not give content to either of Dummett's supposed theses about rules. As Wittgenstein points out,

> It is not clear off-hand what we are to make of the question 'Is $y = x^2$ a formula which determines $y$ for a given value of $x$?' One might address this question to a pupil in order to test whether he understands the use of the word 'to determine'; or it might be a mathematical problem to prove in a particular system that $x$ has only one square. (Wittgenstein 1968, I, §189)

As Wittgenstein observes, in some contexts the question "Is $y = x^2$ a formula which determines $y$ for a given value of $x$?" is a mathematical one. Similarly, in some contexts the question "Do the rules of language system $S$ determine that $s$ is analytic in $S$?" is a mathematical one. In particular, when Carnap asks, "Do the rules of language system $S$ determine that $s$ is analytic in $S$?" he is asking whether, given the arithmetized syntax of $S$, $s$ is an L-consequence in $S$ of the empty set of sentences of $S$. For Carnap, questions about what the rules of a language system $S$ "determine" are just mathematical questions about the L-consequence relation that is specified by an arithmetization of the logical syntax of $S$.[18]

The arithmetical definitions of logical syntax do not settle whether we should accept or reject the indeterminate sentences of a language system. Among these indeterminate sentences are the protocol sentences. Suppose language system S contains the name '$a$' and observational predicate 'is red'. If $S$ is constructed in accordance with Carnap's recommendations, then any two investigators who have decided to use $S$ will agree on how to evaluate the protocol sentence '$a$ is red'. But this does not mean that the rules for using the name '$a$' and observational predicate 'is red', together with facts about the world, 'determine' in some nonmathematical way whether or not the sentence '$a$ is red' is true. For indeterminate sentences, such as '$a$ is red', Carnap does not talk of 'determination' at all. The question which sentences can be treated as protocol sentences of a language system should be addressed and decided by physicists (Carnap

[18] This paragraph differs slightly from the corresponding paragraph of the first published version of this essay, in which I conflated the question of what is provable from the logical syntax of $S$ (as formulated in a metalanguage) with the question of what is an L-consequence of the empty set of premises of $S$. It is the latter question that is key here.

1937, p. 317), whereas the question whether to affirm or deny a particular protocol sentence can only be addressed by investigators who have chosen to *use* the language system *S* that includes that sentence. Neither of these questions is settled by the syntactical rules of *S*.[19]

## 1.8   Carnap's Idealization of Logic

We have seen (in §§1.1–1.5) that if philosophy is replaced by Carnap's logic of science, then there are no contentful philosophical theses, and (in §§1.1–1.5) that the logic of science does not itself presuppose any substantive philosophical or metaphysical theses. If there is anything philosophical about Carnap's logic of science, it is his *motivating attitude* – that there is no point in saying two investigators agree or disagree unless they can be seen to share criteria for evaluating their assertions. To come to terms with Carnap's philosophy we must therefore come to terms with his motivating attitude.

One way to begin is to ask whether Carnap's motivating attitude amounts to an *idealization* of logic that is in tension with our evolving open-ended criteria for evaluating assertions. Consider the following passage from Wittgenstein's *Philosophical Investigations*:

> In philosophy we often *compare* the use of words with games and calculi which have fixed rules, but cannot say that someone who is using language *must* be playing such a game. – But if you say that our languages only *approximate* to such calculi you are standing on the very brink of a misunderstanding. For then it may look as if what we were talking about were an *ideal* language.... – Whereas ... the most that can be said is that we *construct* ideal languages. But here the word 'ideal' is liable to mislead, for it sounds as if these languages were better, more perfect, than our everyday language; and as if it took the logician to show people at last what a proper sentence looked like. (Wittgenstein 1968, I, §81)

Wittgenstein grants that we can construct "calculi which have fixed rules," and he is willing to call these ideal languages. I think he would see Carnap's language systems as calculi of this kind. But Wittgenstein stresses that we

---

[19] Carnap explains that his proposed explications of the term 'observable', which are part of his account of protocol sentences, "belong ... to a biological or psychological theory of language as a kind of the human behavior, and especially as a kind of reaction to observations" (Carnap 1936–7, p. 454). He proposes that we understand 'observable' as follows: "A predicate 'P' of a language L is called observable for an organism (e.g. a person) N, if, for suitable arguments, e.g. 'b', N is able under suitable circumstances to come to a decision with the help of few observations about a full sentence, say 'P(b)', i.e. to a confirmation of either 'P(b)' or '~P(b)'" (Carnap 1936–7, p. 454). On Carnap's treatment of protocol sentences, see Richardson 1996, §3; and Richardson 1998, ch. 9.

"cannot say that someone who is using language *must* be playing such a game." In one sense, Carnap would agree: according to his Principle of Tolerance, everyone is free to use any language he likes. In another sense, however, Carnap would disagree: according to his proposed definition of 'cognitive content,' anyone whose claims are "contentful" (as defined in §1.6 above) expresses those claims by using sentences that can be correlated with indeterminate ("factual") sentences of a constructed language system. Carnap is too sophisticated to insist that there is only *one* way to correlate sentences of a natural language with sentences of a constructed language system. But by proposing that we adopt his definition of 'cognitive content', he expresses his view that constructed language systems are more perfect – more explicit and more exact – than the historically given languages typically used by empirical scientists.

Historically given languages may contain sentences whose methodological role gradually evolves. For instance, without any change in language, a sentence that is at one time treated as a definition may later be regarded as false. This aspect of unreconstructed scientific practice can lead to situations in which investigators take themselves to disagree about some empirical question even though the sentences they use to express their supposed disagreement do not have the same cognitive content (as defined in §1.5). In such situations, Carnap thinks, fruitless controversies like those of traditional metaphysics can arise.

In §79 of the first part of *Philosophical Investigations*, Wittgenstein describes how we may use a name such as 'Moses' to refer to the same person throughout a long period of time even if our 'definition' of the name changes during that time, so that a sentence treated at one time as a definition of 'Moses' is subsequently regarded as false. Wittgenstein also notes that there is a 'fluctuation of scientific definitions: what today counts as an observed concomitant of a phenomenon will tomorrow be used to define it.' He cites changes in definitions to remind us that in practice we do not always regard a change in one of our definitions as a change in subject, as Carnap's syntactical proposals would require us to do.

Hilary Putnam goes further. He claims that fluctuations in scientific definitions show that Carnap's conception of logical syntax is wrong. Here is a characteristic version of Putnam's criticism:

> In Newtonian physics the term momentum was defined as "mass times velocity." ...[But] the principle of Special Relativity would be violated if momentum were *exactly* equal to (rest) mass times velocity.... Can there be a quantity with the properties that (1) it is conserved in elastic collisions, (2) it is closer and closer to "mass times velocity" as the speed becomes

small, and (3) its direction is the direction of motion of the particle? Einstein showed that there is such a quantity, and he (and everyone else) concluded that the quantity *is what momentum really is.* The statement that momentum is exactly equal to mass times velocity was revised. *But this is the statement that was originally a "definition"!* (Putnam 1988, p. 10)

This description of Einstein's revision of the "definition" of 'momentum' conflicts with Carnap's proposals about how to clarify the logic of science.

By Carnap's standards, Einstein's sentence "Momentum is not mass times velocity" is not part of the Newtonian physicists' language system, and so Einstein's assertion of this sentence cannot, strictly speaking, be said to agree or disagree with the Newtonian physicists' assertions of their sentence 'Momentum is mass times velocity'. The reason is that the Newtonian physicists treat their sentence 'Momentum is mass times velocity' as a definition, hence as immune from empirical disconfirmation, whereas Einstein does not. Carnap would correlate the Newtonian physicists' sentence "Momentum is mass times velocity" with an L-valid or P-valid (hence *determinate*) sentence of a language system. We can ask whether it is expedient to adopt the Newtonian language system, within which "Momentum is mass times velocity" is valid, but to reject the system is not to disagree with any of its valid sentences.

Putnam rejects this description of the case. He claims that "the ideas of fixed definitions of terms and analytic truths ... *have no reality for actual scientific practice*" (Putnam 1988, p. 10). As Putnam sees it, the term 'momentum' as it occurs in the Newtonian physicists' sentence "Momentum is mass times velocity" is the same as the term 'momentum' as it occurs in Einstein's sentence "Momentum is not mass times velocity." Hence, in Putnam's view, when Einstein utters his sentence "Momentum is not mass times velocity," what he says conflicts with what Newtonian physicists say when they assert their sentence "Momentum is mass times velocity."

I find this description of the change in definition of 'momentum' compelling, but I do not see why Putnam's appeal to the "reality of actual scientific practice" should convince Carnap to abandon his proposed descriptions of the logical syntax of historically given languages. Carnap stresses that to state the "cognitive content" of a sentence, we must correlate it with a sentence of a language system that we can specify. Carnap's choice of which correlations to accept is guided by his motivating attitude, and so, as Carnap sees it, Newtonian and Einsteinian physicists cannot, strictly speaking, be said to agree or disagree about momentum.

To move beyond this impasse, we must avoid rhetoric about "the reality of actual scientific practice" and try another approach. Let us try comparing the costs and benefits of Carnap's motivating attitude with the costs and benefits of endorsing practices in which investigators regard some terms as unambiguous throughout changes in the methodological role of sentences in which those terms occur.

Without begging the question which rules we "really" follow, we can observe that in practice physicists regard the term 'momentum' as it occurs in the Newtonian physicists' sentence "Momentum is mass times velocity" and the term 'momentum' as it occurs in Einstein's sentence "Momentum is not mass times velocity" as the same. This is displayed by their willingness to take the term 'momentum' as it occurs in the Newtonian physicists' sentence *at face value* – in their willingness to "translate" it homophonically into the language of special relativity, despite fundamental differences between relativity theory and Newtonian physics. In other words, physicists treat 'momentum' as a *transtheoretical* term. Our tendency in scientific investigations to treat some terms as *transtheoretical* embodies what I call *practical realism* – our commitment in practice to the idea that we can be radically wrong.

The main benefit of accepting our practice of treating some terms as transtheoretical is that we can describe and endorse our practical realism. The main cost of accepting this practice – the cost that Carnap would stress – is that there is no guarantee that we will share criteria for recognizing our agreements and resolving our disagreements. We may agree to use a precisely specified method for evaluating a given sentence. But if we accept practical realism, we cannot guarantee that this is the only reasonable method for evaluating the claim we express by affirming the sentence. We may come to accept a different method of evaluating the sentence, without concluding that the sentence expresses a different claim or that the terms that occurred in our earlier uses of the sentence are different from the terms we now use.

In Carnap's view, this cost is too high. He emphasizes that the logical syntax of any sentence or term may be revised "as soon as it seems expedient to do so" (Carnap 1937, p. 318), and so he has no difficulty acknowledging that our methods of evaluating sentences of historically given languages change constantly. But he thinks that we should *not* trust our practice of treating some terms as transtheoretical, because that practice can lead to the kinds of fruitless dispute characteristic of metaphysics.

Although there is no way to *refute* this attitude, one can *invite* those holding it to consider alternatives. I would invite Carnap to trust the open-ended methodology that results from accepting our practice of treating some terms as transtheoretical, despite the risk of falling into fruitless disputes. As we saw above, Carnap accepts that a plausible methodology for empirical science runs the risk of being *inconsistent*. Why is the risk of engaging in fruitless disputes, which scientists have for the most part been able to avoid without using Carnap's methods of specifying logical syntax, so much worse than the risk of inconsistency?

## 1.9 An Internal Tension

Carnap may stick to his methodological scruples despite the invitation to loosen up. There is probably no way to prevent an impasse of alternative attitudes, but it is worth trying to develop a more internal pragmatic criticism. We saw earlier that no system of mathematics can formulate its own logical syntax or prove its own consistency. E. W. Beth has used this technical point to highlight an internal tension in Carnap's approach. Beth observes that if investigators are to use Carnap's methods to specify the logical syntax of Carnap's Language II, for instance, they must take for granted that they share a metalanguage. But, since no amount of purely syntactical or inferential behavior will guarantee that they share a metalanguage, they must trust their practical identifications of shared vocabulary and inference rules. This "mystical" trust seems in tension with Carnap's recommendation that we construct language systems whose logical syntax is fixed and unambiguous. (See Beth 1963, §5f.)

In his reply to Beth, Carnap acknowledges that investigators must trust that they share a metalanguage for use in specifying logical syntax:

> Since the metalanguage ML serves as a means of communication between author and reader or among participants in a discussion, I always presupposed, both in syntax and in semantics, that a fixed interpretation of ML, which is shared by all participants, is given. This interpretation is usually not formulated explicitly; but since ML uses English words, it is assumed that these words are understood in their ordinary senses. The necessity of this presupposition of the common interpreted metalanguage seems to me obvious. (Carnap 1963b, 929)

The necessity that Carnap acknowledges is *practical*: to specify the logical syntax of any language, we must take for granted that we can share some metalanguage whose logical syntax is not itself explicitly specified (though it may be specified, in a different metalanguage, at some later time).

The trouble is that to identify a metalanguage that we share with other investigators, we must trust in our practical identifications of shared vocabulary and inference rules. Such identifications are of a piece with our practical sense of when we agree or disagree with other investigators, and are therefore inextricable from our practical identifications of transtheoretical terms, which Carnap finds methodologically objectionable.

To minimize the danger of falling into fruitless disputes, Carnap recommends that we decide, case by case, which identifications of shared vocabulary and inference rules we trust and which ones we do not:

> It is of course not quite possible to use [an] ordinary language with a perfectly fixed interpretation, because of the inevitable vagueness and ambiguity of ordinary words. Nevertheless it is possible at least to approximate a fixed interpretation ... by a suitable choice of less vague words and by suitable paraphrases. (Carnap 1963b, p. 930)

But Carnap's acknowledgment that we can at best only approximate a fixed interpretation is in tension with his proposal that we avoid vague vocabulary. In effect, he proposes that we use words for which no fixed syntactical rules have been specified to clear up confusions that *stem from* using such words. By clearing up one confusion in this way, we may be just sowing the seeds of another.

This tension is not intolerable, since Carnap does not propose that we simultaneously both use and not use some metalanguage whose logical syntax we have not specified. At worst, Carnap must accept that the work of clarifying the cognitive contents of our assertions is never finished, because it always presupposes the use of some metalanguage or other whose logical syntax we have not (yet) specified. But I think the cost of applying Carnap's method is too high. To apply it, we must make case-by-case stipulations of an arbitrary boundary between trustworthy and untrustworthy identifications of shared vocabulary; all our practical identifications of transtheoretical terms must be deemed untrustworthy, and our corresponding commitment to practical realism must be dismissed as empty and confused. Despite the greater risk that we will be mired in fruitless controversies, I propose that we embrace the open-ended methodology embodied in our practice of using transtheoretical terms.

# CHAPTER 2

# Carnap on Ontology

My central goals here are to show (first) that all of the general existence statements that Rudolf Carnap aims to clarify and defend in his classic essay "Empiricism, Semantics, and Ontology" (Carnap 1950, henceforth ESO) – not only those about abstract objects, such as "There are numbers," but also those about concrete objects, such as "There are physical objects" – are, when viewed in the way Carnap recommends in ESO, analytic (that is, settled solely by the rules of the languages in which they are expressed) and trivially so (that is, derivable from the rules in a few simple steps),[1] and (second) that this interpretive claim is key to understanding how Carnap proposes in ESO to identify and eschew ontological questions.

## 2.1 Background for Reading ESO

Carnap's aim in ESO is to show that "the acceptance of a language referring to abstract entities ... is perfectly compatible with empiricism and strictly scientific thinking" (ESO, p. 206). ESO is addressed to empiricists "who would like to accept abstract entities in their work in mathematics, physics, semantics, or any other field" in the hope that "it may help them to overcome nominalist scruples" (ESO, p. 206).

We know from Carnap's intellectual autobiography (Carnap 1963a, p. 65) that the empiricists he had in mind included W. V. Quine and Nelson Goodman, who, together with Alfred Tarski, asserted in discussions

---

[1] I am not the first to attribute this surprising claim to Carnap; W. V. Quine did so in "On Carnap's Views on Ontology" (Quine 1951). But Quine did not explain or justify the attribution. Perhaps he thought it would be obvious to his and Carnap's intended readers in 1951. The attribution is now widely rejected, however (see Bird 1995, p. 50; Yablo 1998, p. 236; Alspector-Kelly 2001, p. 106; Soames 2009, pp. 428–429; Eklund 2013, p. 245; and Thomasson 2015, pp. 33, 47–48), and therefore calls for a careful explanation and defense.

with Carnap at Harvard in 1940–1941 that a finitistic nominalism (according to which there are only finitely many objects, all of which are physical) is the only scientifically acceptable ontology.[2] This assertion amounts
to a rejection of Carnap's principle of tolerance (first articulated in *The
Logical Syntax of Language* (Carnap 1937, henceforth *Syntax*)), which commits Carnap to being open to specifying and adopting languages (such as
Language I and Language II of *Syntax*) in which one can prove that there
are infinitely many abstract objects. He was troubled both that his fellow
scientifically minded empiricists rejected this aspect of his scientific philosophy, and that they engaged in disputes about existence – including
disputes with Carnap himself – in which "each of the two parties seemed
to criticize the other for using bad metaphysics" (Carnap 1963a, p. 65).
This was especially problematic for Carnap, since it was precisely to avoid
such fruitless disputes that he developed his scientific approach to philosophy. His aim in ESO is to end the fruitless disputes by explaining to
his fellow scientific philosophers, including Quine and Goodman, how it
may *seem* that statements about the existence of abstract objects cannot
be paraphrased into a strictly scientific language (that is, using scientific
terms and logical constructions that Carnap and his fellow scientific philosophers find clear) and are therefore meaningless, when in fact, properly
viewed, they can be so paraphrased and hence are not meaningless.

The scientific philosophy that Carnap takes for granted in ESO rejects
the traditional rationalist idea that there are truths of reason we can know
a priori by means of a special faculty of reason. It also rejects the claim
that logical and mathematical truths are empirical. To reconcile empiricism with these two constraints, in *Syntax* and later works, Carnap formulates logical and mathematical truths so that they are analytic, or true in
virtue of meaning (Carnap 1963a, pp. 46–47, 63–64).

For this purpose Carnap explicates "analytic" relative to what he
calls language systems, which have built-in formation and transformation rules.[3] The formation rules of a language system *LS* together specify
the sentences and terms of *LS*; the transformation rules of *LS* include

---

[2] For a complete German transcription of Carnap's shorthand notes of these conversations, as well as
an English translation of the transcription, see Frost-Arnold 2013.

[3] Here I use the term 'transformation rule' to encompass Tarski-style semantical rules, in addition to
syntactical rules. In *Syntax* Carnap blurred this distinction by defining some syntactical rules – e.g.,
those that settle the logical consequence relation for languages rich enough to express elementary
arithmetic – that amount, in effect, to semantical rules. After Carnap learned of Tarski's method of
defining truth, however, he preferred to define the logical consequence relation for a given language
system in terms of Tarski-style semantical rules.

rules that settle the logical derivability relation between sentences of *LS* and rules that settle the logical consequence relation between sentences of *LS*.

Logical consequence is the relation in terms of which Carnap defines analyticity. Relative to a language system *LS*, a sentence is

analytic (L-true) in *LS* if and only if it is a logical consequence in *LS* of the empty set of sentences of *LS*;

contradictory (L-false) if and only if its negation is analytic (L-true) in *LS*;

synthetic if and only if it is neither L-true nor L-false in *LS*.

Carnap recommends that we construct language systems in terms of rules that we find clear and unproblematic (Carnap 1963b; Ebbs 1997, §§60–61). Some of the analytic sentences of a language system are trivially analytic, in the sense that they can be derived in a few simple steps solely from the rules of the system; other analytic sentences are not trivially analytic. Among the latter, some, such as "There are prime numbers above a hundred," can be derived from the rules of derivation for the language. By Gödel's first incompleteness theorem, however, in any language system *LS* that is rich enough to express elementary arithmetic, there is at least one sentence of *LS* that is L-true but not derivable in *LS*. In such languages L-truth (analyticity) outstrips derivability.[4]

Just as Carnap and other scientific philosophers, including Quine and Goodman, reject any appeal to a faculty of reason to explain why it is reasonable for an inquirer to accept the truths of logic, so they reject any such appeal to explain why it is reasonable to accept that there are abstract objects. Carnap's project in ESO, as I read it, is to explain to his fellow empiricists, or scientific philosophers, how to extend his method of reconciling logic with empiricism to reconcile reference to abstract objects with empiricism. He had done this already in previous work, including *Syntax*, but in ESO he addresses the issue more directly and thoroughly than in any of his previous (or later) writings.

---

[4] Gödel's incompleteness theorems are integral to Carnap's account of analyticity from *Syntax* through all his later works. Theorem 60d.2 of *Syntax* (pp. 221–222) is Carnap's syntactical formulation of Gödel's first incompleteness theorem. Earlier in *Syntax*, immediately after stating Theorem 60C.1 (p. 219), which is, in effect, a syntactical formulation of Tarski's Undefinability Theorem, Carnap notes that 'demonstrable in Language II' can be defined in Language II, but 'analytic in Language II' cannot be defined in Language II.

## 2.2   Linguistic Frameworks and the Internal–External
### Distinction

Carnap writes,

> If someone wishes to speak in his language about a new kind of entities, he
> has to introduce a system of new ways of speaking, subject to new rules; we
> shall call this procedure the construction of a linguistic framework for the
> new entities in question. (ESO, p. 206)

By "language" here Carnap in effect means "language system": he takes
for granted in his discussion in ESO that we have already engaged in a
certain amount of regimentation of what he calls historical given (including natural and scientific) languages (Carnap 1942, §5). Carnap finds it
useful to talk of adding to a given language system LS new formation and
transformation rules that enable one to speak in the resulting extended
language system, $LS'$, about a kind of entity that one could not speak
about in LS. The formation and transformation rules that we need to add
to LS in order to obtain $LS'$ together constitute what he calls a linguistic
framework.[5]

Both our earlier decision to use LS and our decision to extend it to
$LS'$ typically involve what Carnap calls descriptive syntax and semantics (*Syntax* §§24–25; Carnap 1942, §5; Ebbs 1997, pp. 114–121), as well
as decisions to explicate terms that are already in use in natural languages
(Carnap 1947, §2; Lavers 2015).

Carnap writes:

> The two essential steps [in the introduction of a framework] are ... the
> following. First, the introduction [into one's language system] of a general
> term, a predicate of higher level, for the new kind of entities, permitting
> us to say of any particular entity that it belongs to this kind (e.g. "Red is a
> *property*", "Five is a *number*"). Second, the introduction of variables of the
> new type. The new entities are values of these variables; the constants (and
> the closed compound expressions, if any) are substitutable for the variables.
> With the help of the variables, general sentences concerning the new enti-
> ties can be formulated. (ESO, pp. 213–214)

---

[5] My characterization of a linguistic framework accords with Carnap's use of the term 'framework' in
the version of ESO that appears as Supplement A in the second edition of *Meaning and Necessity*
(Carnap 1947). As Carnap explains, in this revised version of ESO, "the term 'framework' is now
used only for the system of linguistic expressions, and not for the system of the entities in question"
(ESO, p. 205, footnote).

It is in terms of the notion of a linguistic framework that Carnap first characterizes his distinction between internal and external questions, as follows:

> We must distinguish two kinds of questions of existence: first, questions of the existence of certain entities of the new kind *within the framework*; we call them *internal questions*; and second, questions concerning the existence or reality *of the system of entities as a whole*, called *external questions*. Internal questions and possible answers to them are formulated with the help of the new forms of expressions. (ESO, p. 206)

This is just a preliminary sketch, of course; a large part of ESO is spent just filling in the sketch.

I shall argue that a full explanation of what Carnap means by calling a question "external" depends on a proper understanding of a crucial distinction he draws between two different types of *internal* questions – i.e., two types of questions that can be formulated in a linguistic framework. Carnap explains how to draw his crucial distinction between types of internal questions for several different linguistic frameworks, including two paradigmatic frameworks that I shall focus on here: the number framework and the physical object (thing) framework.[6]

I'll start with the number framework, since Carnap's characterization of it is sharper and more developed than his characterization of the physical object framework. He writes:

> The framework for [numbers] ... is constructed by introducing into the language new expressions with suitable rules: (1) numerals like "five" and sentence forms like "there are five books on the table"; (2) the general term "number" for the new entities, and sentence forms like "five is a number": (3) expressions for properties of numbers (e.g., "odd", "prime"), relations (e.g., "greater than"), and functions (e.g., "plus"), and sentence forms like "two plus three is five"; (4) numerical variables ("$m$", "$n$", etc.) and quantifiers for universal sentences ("for every $n$, ...") and existential sentences ("there is an $n$ such that ...") with the customary deductive rules. (ESO, p. 208)

In the framework for numbers one can ask, "Is there a prime number greater than a hundred?" and "Are there numbers?" The answers to both of these internal questions are found "by logical analysis based on the rules for the new expressions." The answer to the second of these internal questions is different from the answer to the first, however, since

---

[6] In ESO Carnap also, though more briefly, discusses linguistic frameworks for propositions, thing (physical object) properties, integers and rational numbers, real numbers, and space–time points.

["there are numbers", or "There is an $n$ such that $n$ is a number"] follows from the analytic statement "five is a number" and is therefore itself analytic. Moreover, it is rather trivial (in contradistinction to a statement like "There is a prime number greater than a million", which is likewise analytic but far from trivial), because it does not say more than that the new system is not empty; but this is immediately seen from the rule which states that words like "five" are substitutable for the new variables. (ESO, p. 209)

I assume that by "the new system is not empty" Carnap means that the domain of its variables is not empty. Given this assumption, I take Carnap in this passage to be making the following two claims:

(i) 'Five is a number' is analytic.
(ii) From the transformation rule that states that words like 'five' are substitutable for the number variables, we can see immediately that the domain of entities over which the number variables range is not empty.

A central goal of my reading of ESO is to explain why Carnap accepts these two claims; I will discuss each of them in §2.3.

Carnap's observation that "There are numbers" is a trivially analytic statement of the framework of numbers is central to his characterization of the kinds of questions about the existence of numbers that he regards as problematic and calls "external." He reasons as follows:

[Since the answer to the question "Are there numbers?" in the internal sense is trivially analytic] nobody who meant the question "Are there numbers?" in the internal sense would either assert or even seriously consider a negative answer. This makes it plausible to assume that those philosophers who treat the question of the existence of numbers as a serious philosophical problem and offer lengthy arguments on either side, do not have in mind the internal question. And, indeed, if we were to ask them: "Do you mean the question as to whether the framework of numbers, *if* we were to accept it, would be found to be empty or not?", they would probably reply: "Not at all; we mean a question prior to the acceptance of the new framework". They might try to explain what they mean by saying that it is a question of the ontological status of numbers; the question whether or not numbers have a certain metaphysical characteristic called reality...or subsistence or status of "independent entities". Unfortunately, these philosophers have so far not given a formulation of their question in terms of the common scientific language. Therefore our judgment must be that they have not succeeded in giving to the external question and to the possible answers any cognitive content. Unless and until they supply a clear cognitive interpretation, we are justified in our suspicion that their question is a pseudo-question, that is, one disguised in the form of a theoretical question while in fact it is non-theoretical; in the present case it is the practical problem whether or not to incorporate into the language the new linguistic forms which constitute the framework of numbers. (ESO, p. 209)

In this passage Carnap identifies the external question "Are there numbers?" in part by noting that a person who voices it should not be interpreted as asking the internal question "Are there numbers?" since "nobody who meant the question 'Are there numbers?' in the internal sense would either assert or even seriously consider a negative answer." This method of identifying external questions is central to the reading of ESO that I shall develop below.

Carnap's treatment of existence questions about physical objects is parallel in several ways to his treatment of questions about numbers. He notes that in the physical object language, one can ask such questions as "Is there a white piece of paper on my desk?" and "Are unicorns real or imaginary?" where the second of these questions amounts to "Are there any unicorns?" or (when rewritten in a linguistic framework for speaking about physical objects) "Is there an $x$ such that $x$ is a unicorn?" About internal questions of this first type, which parallel the question "Are there prime numbers greater than a hundred" in the framework for numbers, Carnap writes:

The concept of reality [i.e. existence] occurring in these internal questions is an empirical, scientific, non-metaphysical concept. To recognize something as a real thing or event means to succeed in incorporating it into the system of things at a particular space-time position so that it fits together with the other things recognized as real, according to the rules of the framework. (ESO, p. 207)

Although Carnap speaks very loosely here, I think it is clear that his main point is that the answers to internal questions such as "Is there a white piece of paper on my desk?" and "Are there any unicorns?" are settled by the rules of the framework, together with sentences in the framework that express the results of empirical observations.

It is less clear, and one of the issues I shall try to resolve, whether there is an internal question in the physical object framework that corresponds with the internal question "Are there numbers?" in the number framework. Carnap writes:

If someone decides to accept the thing [physical object] language, there is no objection against saying that he has accepted the world of things [physical objects]. But this must not be interpreted as if it meant his acceptance of a belief in the reality of the thing world; there is no such belief or assertion or assumption, because it is not a theoretical assumption. To accept the thing world means nothing more than to accept a certain form of language, in other words, to accept rules for forming statements and for testing, accepting, or rejecting them. (ESO, p. 208)

In what sense is "acceptance of the thing world" not a theoretical assumption, in Carnap's view? The answer, I suggest, is that such acceptance is an analytic consequence of adopting what Carnap calls the thing language. As I shall explain below, in a customary thing language one can form the internal statement "There are physical objects," or, more explicitly, "There is an $x$ such that $x$ is a physical object." It appears that in the above passage Carnap is saying that to accept a customary thing language is thereby to commit oneself to accepting the internal statement "There are physical objects." This may seem to involve "acceptance of a belief in the reality of the thing world" in a metaphysical sense characteristic of traditional ontology, but, in Carnap's view, it does not.[7]

Carnap thinks that in a customary thing language the internal statement "There are physical objects" is different in kind from a statement such as "There is a piece of paper on my desk." I suggest we explain the difference he has in mind by developing parallels with what he says about the analogous statements in the framework for numbers, as follows:

If someone decides to accept the physical object language, there is no objection against saying that he has thereby accepted that there are physical objects. Therefore, nobody who meant the question "Are there physical objects?" in the internal sense would either assert or even seriously consider a negative answer. This makes it plausible to assume that those philosophers who treat the question of the existence of physical objects as a serious philosophical problem and offer lengthy arguments on either side, do

---

[7] My interpretation of Carnap's claim that "a belief in the reality of the thing world ... is not a theoretical assumption," is partly guided by the following passage from Carnap's paper "W. V. Quine on Logical Truth" (1963c):

[Philosophical principles or doctrines, including] proposals for certain explications (often not stated explicitly) and of certain assertions which, on the basis of these explications, are analytic....are sometimes called theories; however, it might be better not to use the term "theory" in this context, in order to avoid the misunderstanding that such doctrines are similar to scientific, empirical doctrines. (Carnap, 1963c, p. 917)

On my reading, these cautionary remarks about how to use the word 'theory' imply parallel cautionary remarks about how to use the word 'theoretical'. When Carnap writes that "there is no such belief or assertion or assumption [as] a belief in [or assertion of assumption of] the reality of the thing world, because it is not a theoretical assumption," he is rejecting the traditional philosophical view that such a belief or assertion or assumption is synthetic yet knowable only by *a priori* methods. Thus his rejection of "a belief in the reality of the thing world" as "non-theoretical" is an application of the general strategy for rejecting metaphysical questions that I explain in Chapter 1, §1.5.

not have in mind the internal question. And, indeed, if we were to ask them: "Do you mean the question as to whether the framework of physical objects, *if* we were to accept it, would be found to be empty or not?" they would probably reply: "Not at all; we mean a question prior to the acceptance of the new framework." They might try to explain what they mean by saying that it is a question of the ontological status of physical objects; the question whether or not physical objects have a certain metaphysical characteristic called reality ... or subsistence or status of "independent entities." Unfortunately, these philosophers have so far not given a formulation of their question in terms of the common scientific language. Therefore our judgment must be that they have not succeeded in giving to the external question and to the possible answers any cognitive content. Unless and until they supply a clear cognitive interpretation, we are justified in our suspicion that their question is a pseudoquestion, that is, one disguised in the form of a theoretical question while in fact it is non theoretical; in the present case it is the practical problem whether or not to incorporate into the language the new linguistic forms which constitute the framework of physical objects.

This way of reading Carnap's remarks about the number framework and the physical object framework sets up the following parallels between the various types of questions that Carnap distinguishes in the two frameworks:

|  | Number Framework | Physical Object Framework |
| --- | --- | --- |
| Internal Qs |  |  |
| Nontrivial | Are there prime numbers above a hundred? | Is there a piece of paper on my desk? Are there unicorns? |
| Trivial | Are there numbers? | Are there physical objects? |
| External Qs | ARE THERE NUMBERS? | ARE THERE PHYSICAL OBJECTS? |

Carnap reasons that since the answers to "Are there numbers?" and "Are there physical objects?" understood as questions internal to the number framework and the physical object framework, respectively, are trivial, those who try to use the sentences "Are there numbers?" and "Are there physical objects?" to raise difficult theoretical questions and to offer lengthy arguments on either side do not have the internal questions in mind. He argues that such philosophers "have so far not given a formulation of their question in terms of the common scientific language" and concludes that "they have not succeeded in giving to the external question and to the possible answers any cognitive content." Finally, he offers an interpretation of the questions that does not ascribe any cognitive (truth-evaluable) content to them. The puzzling external questions, ARE THERE NUMBERS? and ARE THERE PHYSICAL OBJECTS?, are best understood, he thinks, as practical questions about whether or not to incorporate into one's language the framework of numbers and of physical objects, respectively.

## 2.3   Universal Words, Proper Names, Designation, and Analyticity

I shall argue that when the number framework and the physical object framework are constructed in the way that Carnap recommends in ESO, the trivial internal questions have trivially analytic or contradictory answers, and the nontrivial internal questions have either nontrivial analytic or contradictory answers, or have synthetic answers, depending on the framework in which the questions are posed. In particular, when formulated in the way that Carnap recommends, the questions "Are there numbers?" and "Are there physical objects?" have trivially analytic answers.[8]

This claim, especially the claim that the answer to the question "Are there physical objects?" is trivially analytic, is surprising, to say the least. Many philosophers regard it as obviously false. Nevertheless, as I shall now argue, the same pattern of reasoning that Carnap uses to show that 'There are numbers' is trivially analytic in the number framework, including the reasoning that secures the two main claims (i) and (ii) that I identified above, also establishes that 'There are physical objects' is trivially analytic in the physical object framework.

Consider first why Carnap accepts (i), according to which 'Five is a number' is trivially analytic in the number framework. In *Meaning and Necessity* (Carnap 1947), published just three years before ESO, Carnap writes:

> Generally speaking, if a language (of ordinary structure) contains certain variables, then we can define in it a designator for the range of values of those variables. In the present case, the definition is: ... "'Number $(m)$' for '$m = m$'". (In the definiens, any matrix '$...m...$' may be used which is

---

[8] For a similar, though less detailed, reading of ESO, see Quine 1951, pp. 68–69. Bird 1995 takes Quine to conflate what Quine calls category questions, which are the trivially analytic or contradictory general existence questions of a given framework, with what Carnap calls external questions. I read Quine differently. He writes, "The internal questions comprise ... the category questions when these are construed as treated with the adopted language as questions having trivially analytic or contradictory answers" (Quine 1951, p. 69). This is not a modification of Quine's reading, as Bird suggests (Bird 1995, p. 50). What Quine earlier calls "Carnap's dichotomy of questions of existence" is Carnap's dichotomy of *internal* questions of existence. Quine's word "dichotomy" refers back to the bottom of page 67, where he is clearly describing the two kinds of internal questions.

L-universal, that is, such that '$(m)(\ldots m \ldots)$' is L-true.) … once you admit certain variables, you are bound to admit the corresponding universal concept. (Carnap 1947, p. 44, with Carnap's square brackets changed to parentheses)

In this passage Carnap affirms the following general principle:

> (I)  In any linguistic framework that contains variables of a given style – variables for numbers or physical objects, for instance – there is a corresponding universal word, such that it is L-true (analytic) that this universal word holds of every entity over which the variables of the framework range.

A universal word of a framework expresses what Carnap calls the universal concept for the framework. In the number framework, for example, the universal word 'number $(m)$' may be defined as '$m = m$', and '$(m)(m = m)$' is L-true. Carnap's claim (i) is justified by principle (I), on the assumption, which I shall examine below, that the names in the number framework designate numbers. For in the number framework, which contains the numeral 'Five', 'Five is a number' amounts to 'Five = Five', which (by the customary rules for identity) is L-true (analytic).

Although in ESO Carnap does not restate principle (I), ESO is addressed to readers who know principle (I) and understand how it follows from basic truths of mathematical logic. This is clear from the fact that in Carnap's most explicit justification in ESO of the claim that 'Five is a number' is analytic, he supposes that

> our language L contains the forms of expressions which we have called the framework of numbers, in particular, numerical variables and the general term "number",

and reasons that

> if L contains these forms, the following is an analytic statement in L: (b) "Five is a number". (ESO, p. 217)

The antecedent of this conditional – "L contains these forms" – provides a reason for accepting the consequent of the conditional – "'Five is a number' is analytic" – only if the universal words of a given linguistic framework are L-true (analytic) of every entity over which the variables of the framework range – or, in other words, only if principle (I) is true.

The same kind of reasoning shows that if our language system contains the physical object framework (including variables that range over physical objects, names of physical objects, and the universal word 'physical object $(x)$', defined as '$x = x$', where '$x$' is the style of variables used for physical objects in the framework), then if '$n$' is one of the names of the physical object framework, the sentence 'physical object ($n$)' is analytic in the framework. In particular, if 'Fido' is a name in the framework for physical objects, as Carnap assumes (see ESO, p. 217), then 'Fido is a physical object' is defined as 'Fido = Fido', which is L-true (analytic) in the framework.

Although in ESO Carnap does not explicitly apply this pattern of reasoning to names of physical objects, in *Syntax* Carnap provides examples of universal words, including 'thing', understood as 'physical object', and writes:

> In the word-series 'dog', 'animal', 'living creature', 'thing', every word is a more comprehensive predicate than the previous one, but only the last is a universal predicate [in a language system with a separate style of variables for things]. In the corresponding series of sentences, 'Caro is a dog', '... is an animal', '... is a living creature', 'Caro is a thing', the [empirical] content is successively diminished. But the final sentence is fundamentally different from the preceding ones, in that ... it is analytic. (*Syntax*, p. 293)

One might think that between the time he wrote *Syntax* and ESO, Carnap changed his views on this point. But the passage I quoted above from *Meaning and Necessity*, published just three years before ESO, commits Carnap to concluding that 'Caro is a thing' is analytic in thing framework, for essentially the same reasons that he presents in the above passage in *Syntax*. His views about the resources for constructing language systems changed in the period between the publication of *Syntax* and the publication of ESO, but I know of no textual evidence that Carnap changed his mind about whether a sentence such as 'Caro is a thing', where 'thing' is a universal word, and hence exhausts the range of the variables for which 'Caro' may be substituted, is analytic.

The case in favor of interpreting Carnap's reasoning in ESO as dependent on principle (I) looks even stronger when one notices that in the paragraph immediately following the above-quoted passage about universal concepts from *Meaning and Necessity*, Carnap writes:

> In my view ... the accusation of an absolutist metaphysics or of illegitimate hypostatizations with respect to a certain kind of entities, say propositions, cannot be made against an author, merely on the basis of the fact that he uses variables of the type in question (e.g. '$p$', etc.) and the corresponding universal word ('proposition'); it must be based, instead, on an analysis of the statements or pseudo-statements which he makes with the help of those signs. (Carnap 1947, p. 44)

Indeed, the entire section (§10) of *Meaning and Necessity* in which these passages occur is clearly a preliminary sketch of Carnap's argument in ESO that to admit into one's language variables of a certain style, such as numerical variables, and the corresponding universal word, such as 'number', is not thereby to engage in "absolutist metaphysics" or "illegitimate hypostatizations" with respect to the values of such variables.

Let us now consider Carnap's claim (ii), namely, that from the transformation rule that states that words like 'five' are substitutable for the number variables, we can see immediately that the domain of entities over which the number variables range is not empty. This claim also follows from a more general principle that Carnap accepts. Before stating the principle I need to introduce some terminology. I shall call a linguistic framework *customary* if it contains names that are substitutable for the variables of the framework, and there is a rule of EG that allows us to infer '$(\exists v)(Fv)$' from '$Fn$', where 'F' is a predicate of the framework, '$v$' is a variable in the framework, and '$n$' is a name of the framework that one can substitute for '$v$'. With this terminology, we can see that claim (ii) follows from

> (II) From a transformation rule that states that the names of a customary linguistic framework are substitutable for the variables of that framework, we can trivially infer that the domain of entities over which the variables range is not empty.

The reasoning is as follows. In a customary framework the sentence that paraphrases "the domain of entities over which the variables range is not empty" has the form '$(\exists v)(Uv)$', where 'U' is a universal word of the framework, such as 'number' or 'physical object'. As we saw above, by principle (I), for any name '$n$' of a customary framework, '$Un$' is analytic. Since '$(\exists v)(Uv)$' is a logical consequence of '$Un$', and a logical consequence of an analytic sentence is analytic, '$(\exists v)(Uv)$' is analytic. In particular, since 'There are numbers' is a logical consequence of 'Five is a number', and 'Five is a number' is analytic, 'There are numbers' is analytic. For exactly the same reasons, since 'There are physical objects' is a logical consequence of 'Fido is a physical object', and 'Fido is a physical object' is analytic, 'There are physical objects' is analytic. In just this way, by applications of principles (I) and (II) we can trivially infer that the domain of entities over which the variables of the number framework and the physical object framework, respectively, range is not empty.

To apply principles (I) and (II) one must assume that the names of a customary framework designate entities in the domain over which the

variables for which they can be substituted range. One might think, however, that it cannot be analytic according to Carnap that a name '$n$' designates an entity in the domain over which the variables it can be substituted
for range, and infer from this supposition that the step from '$Un$' to '$(\exists v)$
$(Uv)$' cannot be analytic according to Carnap, regardless of whether the
framework is logical or factual.

The inference is fine, but the thought is mistaken. According to Carnap,
the extensions of the names of a framework are values of the variables
for which the names can be substituted (Carnap 1947, §§9–10), and the
rules of designation for the names of a framework settle the extensions of
those names (Carnap 1947, pp. 170–171). Carnap therefore explains the
step from '$Un$' to '$(\exists v)(Uv)$' in a given framework by including rules of
designation for the names of the framework. He writes:

> Generally speaking, any expression of the form "'…' designates …" is an
> analytic statement provided the term "…" is a constant in an accepted
> framework. If the latter condition is not fulfilled, the expression is not a
> statement. (ESO, p. 217)

This implies that if "Fido" is a constant in an accepted customary framework for physical objects, as Carnap assumes in ESO (see p. 217), then

(a) 'Fido' designates Fido.

is analytic in that framework. As we saw above, his principle (I) implies
that

(b) Fido is a physical object.

or, equivalently, 'Fido = Fido', is analytic in the framework. And from
(a) and (b) we may infer that

(c) 'Fido' designates a physical object.

Since (a) and (b) are analytic, so is (c). The supposed concern that the
name 'Fido' fails to designate a physical object, so that (c) is false, is ruled
out by the analyticity of (c), which is demonstrated by the fact that (c) follows from the analytic truths (a) and (b).

One might suppose that by a series of similar inferences one could also
get the result that, for instance, ' "Pegasus" designates Pegasus', 'Pegasus
is a physical object', and ' "Pegasus" designates a physical object' are also
analytic in such a physical object framework.[9]

This supposition overlooks the fact that for Carnap the adoption of
a physical object framework is a regimentation of natural language

[9] I thank Stephen Yablo for urging me to address this issue.

sentences that includes an explication of the ordinary-language grammatical category of a proper name. Carnap's semantic principle that *if the
term '...' is not a constant in an accepted framework, the expression ' "..."
designates ...' is not a statement* (ESO, p. 217) is a constraint on our explication of proper names. Given this principle, no one who is aware of the
mythological origins of the term 'Pegasus' would list 'Pegasus' among the
names of a customary framework for speaking about physical objects. And
if 'Pegasus' is not on the list of the names of an accepted customary framework for physical objects, then by Carnap's semantic principle ' "Pegasus"
designates Pegasus' is not among the rules of designation for any such
framework and hence cannot be used to derive ' "Pegasus" designates a
physical object'.[10]

A person who is unaware of the mythical origins of the name 'Pegasus'
might list 'Pegasus' among the names of a customary framework and use
the orthographic word form 'physical object' as a universal word of the
framework. Suppose Alex constructs such a framework. In his own metalanguage for the framework, Alex can then lay down, as analytic, sentences of the form '"Pegasus" designates Pegasus' and 'Pegasus is a physical
object'. Alex's acceptance of these sentences does not yet tell us what
he means by them, however. In particular, the fact that Alex's sentence
'"Pegasus" designates a physical object' is analytic for the framework he
has constructed does not by itself establish either that his framework is a
physical object framework or that in his framework, 'Pegasus' designates a
physical object. To understand Alex's sentences, we need to translate them
into a language we can use. Since sentences of the form '"Pegasus" designates Pegasus' and 'Pegasus is a physical object' are analytic in Alex's
language (i.e., the language he uses to state the semantical rules for the
linguistic framework he has constructed), to understand these semantical
sentences we would need to identify analytic sentences of our language

[10] This still leaves us with the option of explicating a name such as "Pegasus" as a descriptive singular
term, such as "the winged horse that was captured by Bellerophon." In *Meaning and Necessity*,
Carnap presents a neo-Fregean account of descriptive singular terms, according to which "the rules
of a language system should be constructed in such a way that every description has a descriptum"
(Carnap 1947, p. 35). The idea is that we can decide in advance what object a given descriptive
singular term shall refer to, if, as it may be, nothing satisfies the description we ordinarily associate
with that term. For instance, we may decide that if nothing satisfies the descriptive singular term
'the winged horse that was captured by Bellerophon', then the term refers to Secretariat. (Carnap
acknowledges that any such choice will be "more or less arbitrary," but, following Frege, he thinks
"the disadvantage seems small in comparison with the gain in simplicity for the rules of the system"
(Carnap 1947, p. 35).) When "the winged horse that was captured by Bellerophon" is explicated in
this way, then since there is no winged horse that Bellerophon captured, the singular descriptive
term 'the winged horse that was captured by Bellerophon' designates Secretariat, a physical object.

that translate them. We will not find such sentences until we realize that the universal word 'physical object' that Alex uses in his semantical sentences is not true of all and only physical objects, in the sense of 'physical object' that is defined as '$x = x$' in our customary physical object framework, but applies, instead, to all and only the entities (whatever they are) in the domain of the framework Alex has constructed. Moreover, we might not be able to translate Alex's sentences into our own language, as it now exists. It may be that the only way for us to understand the universal word 'physical object' that Alex uses in his semantical sentences is to learn his language.

Carnap's semantic and syntactical rules do not guarantee, of course, that a framework we accept at a given time will prove useful to us. No matter how useful a given physical object framework is to us now, we may later come to think that it includes among its names some singular terms that we no longer wish to treat as names of physical objects. In such cases, we would no longer accept the old framework; but would replace it with another framework that we find more suited to our purposes.

## 2.4  Carnap's Method of Identifying External Questions

I conclude that when "There are numbers" and "There are physical objects" are formulated in the way Carnap recommends in ESO, they are trivially analytic, and their negations are trivially contradictory. Moreover, in the number framework and the physical object framework there are existence statements that are neither trivially analytic nor trivially contradictory; these existence statements require a certain amount of proof (in the number framework) or investigation (in the physical object framework) to answer. I shall call Carnap's distinction between these two kinds of existence statements in a given framework his trivial/nontrivial distinction, for short.

Carnap's trivial/nontrivial distinction is central to the method for identifying metaphysical questions about existence that Carnap summarizes in the long passage from ESO, p. 209, that I quoted in §2.2. This passage makes sense in light of the following four observations.

**First Observation** If an existence question, such as "Are there prime numbers above a hundred?" or "Are there elephants in Alaska?" is formulated so that its answer is not trivially analytic or contradictory – i.e., not derivable in a few simple steps solely from the rules for evaluating sentences of the framework in which it is raised – we can understand how someone could be in doubt about how to answer it, and, also, how the question could at least in principle be answered by applying the relevant rules of the framework (whether they be rules of proof, or rules of verification by accepted protocol sentences of the language).

**Second Observation** If a general existence question, such "Are there numbers?" or "Are there physical objects?" is formulated so that its answer is trivially analytic or contradictory – i.e., derivable in a few simple steps solely from the rules for evaluating sentences of the framework in which it is raised – we cannot understand how anyone could be in any doubt about how to answer it.

**Third Observation** If for the reasons described in the second observation we do not understand how a person who knows the rules of a given language system could be in any doubt about how to answer a given question formulated in that language system, then we should not interpret the person as asking that question, but should instead try to find another way of interpreting the question.

**Fourth Observation** If, as Carnap and his fellow scientific philosophers assume, every meaningful question can be formulated in some language system or other, then "unless and until [a person whose question we are trying to interpret for the reasons described in the third observation] suppl[ies] a clear cognitive interpretation of [his] question, we are justified in our suspicion that [his] question is a pseudo-question, that is, one disguised in the form of a theoretical question while in fact it is nontheoretical" (ESO, p. 209).

These observations are expressed in terms of, and therefore presuppose, Carnap's trivial/nontrivial distinction. The first observation, which Carnap leaves implicit in his remarks in ESO, is that if a question has a nontrivial answer, we have no difficulty understanding how a person could be in doubt about its answer. It is only when a person seems to be raising a question whose answer is trivially analytic or contradictory that Carnap's second and third observations about interpretation apply, so that we may end up classifying the person's question as "external." In traditional philosophy, and sometimes even in scientific philosophy, because of confusion, a person may voice such questions as "Are there numbers?" and "Are there physical objects?" while refusing to accept any trivially analytic answers to the questions. In such cases, according to the second and third observations, we should not interpret the person as asking meaningful existence questions with trivially analytic answers, but should instead try to find another way of interpreting the questions. It is natural to try to understand the person as asking general existence questions that are similar in form to internal general existence questions, but independent of any linguistic framework. By the fourth observation, such questions, which seem to concern "the existence or reality *of the system of entities as a whole*," are what Carnap calls "*external questions*" (p. 206), or, in other words, nontheoretical, pseudoquestions that are disguised in the form of theoretical questions (ESO, p. 209).[11]

## 2.5  The Scope and Limits of Carnap's Method

Carnap's method for identifying metaphysical statements about existence can be applied relative to any language system in which one can draw his trivial/nontrivial distinction. We can extend his method to language systems he does not discuss in ESO if we can draw his trivial/nontrivial distinction for such language systems.

In ESO Carnap draws his trivial/nontrivial distinction only for linguistic frameworks with special types of variables and names, such as a framework for numbers and a distinct framework for physical objects, as described above. It is relative to a given framework of this kind that he relies on his principles (I) and (II) to show that general existence claims of the form '$(\exists v)(Uv)$', where 'U' is a universal word of the framework, such as 'number' or 'physical object', are trivially analytic. He does not explain how one might draw the trivial/nontrivial distinction for existence claims about numbers or physical objects in a language system with a single style of variables that range over both numbers and physical objects, for instance. Can the trivial/nontrivial distinction be drawn for such language systems?

---

[11]  In a letter to Carnap written in 1947, before Quine finally gave up the analytic–synthetic distinction, Quine proposed a similar method of identifying ontological truths and explaining why debate about them is "hard to settle":

> Perhaps a typical feature of ontological truths is that they are analytic statements of a kind which would be too trivial to invite assertion or dispute except for doubt or disagreement as to adoption or retention of special features of the language on which their truth depends. And such disagreements are hard to settle simply because the basic features of the language or languages in which the dispute takes place are themselves at stake, depriving the disputants of a fixed medium of discussion. (Quine, Letter to Carnap, May 1, 1947, in Creath 1990, pp. 409–412; quotation from p. 410)

> This suggestion may have played a role in Carnap's development of the method of identifying "external" questions that he presents in ESO.

The answer is "yes."[12] Suppose that *LS* is a language system with separate variables and names for numbers and physical objects, so that by principles (I) and (II), 'Five is a number' and 'Fido is a physical object' are trivially analytic sentences. Then one can construct a language *LS'*, with a single style of variables and names, whose transformation rules list 'Five is a number' and 'Fido is a physical object' as primitive meaning postulates (Carnap 1952b), so that 'Five is a number' and 'Fido is a physical object' are trivially analytic in *LS'*. One can then apply the transformation rule EG of *LS'* to derive 'There are numbers' and 'There are physical objects', thereby showing that these two sentences are also both trivially analytic in *LS'*. Similarly, one can construct *LS'* so that corresponding to "There are primes above a hundred" and "There are elephants in Alaska" of *LS*, there are sentences of *LS'* that are not trivially analytic, but that require proof and empirical inquiry, respectively, to establish.

In general, for any language system *LS* with separate styles of variables and names for different types of entities, such as numbers or physical objects, one can construct another language system *LS'* with only one type of variable in which there are (1) trivially analytic or contradictory sentences that translate the trivially analytic or contradictory sentences of *LS*, and (2) sentences of *LS'* that translate those sentences of *LS* that require a certain amount of proof or investigation to evaluate. Sentences of type (2) will be either nontrivially analytic or synthetic according as they translate nontrivially analytic or synthetic sentences of *LS*. In particular, one can construct *LS'* so that the trivially analytic sentences of *LS* have as counterparts equally trivially analytic sentences of *LS'*.

In short, Carnap does not need to draw the trivial/nontrivial distinction for statements about numbers or physical objects by introducing language systems with separate variables and names for numbers and physical objects. The adoption of such language systems is therefore not indispensable to his method in ESO.[13]

---

[12] Quine 1951, p. 71, makes the same claim but does not explain it.
[13] Why then does Carnap express his central points in ESO in terms of language systems with separate variables and names for numbers and physical objects? Part of the answer, I believe, is that Carnap thinks we are independently motivated to adopt language systems with segregated variables by our desire to regiment language so that what Carnap takes to be garden-variety meaningless strings of letters and spaces, such as 'Fido is divisible by 3' and 'Nine is a dog', which might by our ordinary grammatical standards be classified as sentences, are not sentences of the language system we use to clarify the cognitive contents of sentences of English (Carnap 1932, pp. 67–68). He also thinks that use of a single style of variables in set theory tends to increase the risk that one's set theory is inconsistent. (See Carnap, Letter to Quine, April 13, 1947, in Creath 1990, pp. 405–407; quotation from p. 406; cited in Lavers 2015.) If we are already committed to adopting language systems with separate variables and names for different types of entities, as Carnap thinks, then

What is indispensable to his method in ESO is that we be willing to paraphrase the languages we use so that under the paraphrases, it is analytic that the domain (or domains) of the variables we use is (or are) not empty. He does not claim that we must adopt such paraphrases, only that we may do so if we wish. His principle of tolerance, to which he was committed as early as 1934, in *Syntax*, and to the end of his career, rules out any stronger claim.

Carnap himself sometimes preferred to use what he called "coordinate languages," in which "the form of an individual expression indicates the position of that individual in the basic ordering system" (Carnap 1958, p. 161; see also *Syntax*, pp. 12, 45; Carnap 1947, pp. 57, 79; ESO, pp. 212–213). In such languages, Carnap explains, "Numerical expressions or $n$-tuples of such (when the basic ordering is $n$-dimensional) appear as individual expressions" (Carnap 1958, p. 161). In a language system with a spatiotemporal coordinate system for physics, which includes variables with space–time points as values and quadruples of real numbers that designate space–time points, by Carnap's principles (I) and (II) we may trivially infer that there are space–time points (ESO, p. 213).[14] Whether we can thereby *also* infer the existence of physical objects (things) depends on how we choose to explicate the term "physical object".[15] If we explicate "physical object" as "class of space–time points" (Carnap 1958, p. 161), then in a customary spatiotemporal coordinate system for physics it is trivially analytic that there are physical objects. If we explicate "physical object" so that "whether [some physical object] (matter or an electro-magnetic field) is to be found at a particular position is expressed by the fact that at the position in question the value of the density – or of the field vector, respectively – is not zero," however, then "whether anything [i.e., any physical object] at all exists – that is to say, whether there is a non-trivially

it is natural to prefer such language systems when we try to formulate and evaluate the puzzling questions about existence that philosophers raise. This assumption looks even more attractive from Carnap's point of view when one realizes how many artificial stipulations one needs to make in order to construct a language system without segregated variables in which such sentences as 'There are numbers' and 'There are physical objects' are analytic.

[14] Bird 1995 claims that in what Carnap calls "the spatio-temporal coordinate system for physics" (ESO, p. 212), only the existence of quadruples of real numbers, not the existence of space–time points, is analytic. In ESO, p. 213, however, Carnap writes: "A question like 'Are there (really) space-time points?' is ambiguous. It may be meant as an internal question; then the affirmative answer is, of course, analytic and trivial" (ESO, p. 212). The internal question concerns the existence of space–time points, not (as Bird claims) the existence of quadruples of numbers, which are designations of space–time points, not space–time points themselves.

[15] I thank Richard Creath for pointing this out to me.

occupied position can only be expressed by means of synthetic sentences" (*Syntax*, p. 141).[16]

One might think this poses a problem for my reading of Carnap's project in ESO – that the reading is in trouble if there are *any* language systems in which "There are physical objects" is not analytic. Recall, however, that Carnap's aim in ESO is to show that "the acceptance of a language referring to abstract entities … is perfectly compatible with empiricism and strictly scientific thinking" (ESO, p. 206). It is addressed to empiricists "who would like to accept abstract entities in their work in mathematics, physics, semantics, or any other field" in the hope that "it may help them to overcome nominalist scruples" (ESO, p. 206). Carnap's central points in ESO are that some general existence statements, including both "There are numbers," which appear to some empiricists to be meaningless, and "There are physical objects," which empiricists find unproblematic, can be expressed in customary linguistic frameworks, and, thus expressed, are trivially analytic, and that empiricists who challenge such internal existence statements unwittingly ask an "external" question – i.e., a pseudoquestion that cannot be expressed in any language system, and therefore has no cognitive (truth-evaluable) content. To make these points Carnap does not, and need not, explain the contrast between "external" and "internal" for all possible linguistic frameworks.

## 2.6   Apriority and the Analytic–Synthetic Distinction

Perhaps the most basic source of resistance to the view I attribute to Carnap is the assumption that it is just plain wrong, even repugnant, to say that "There are physical objects" is analytic. Don't we know both that the truth of an analytic sentence can be known a priori and that the truth of "There are physical objects" cannot be known a priori? If so, then we may immediately infer that "There are physical objects" is not analytic.

The problem with this objection is that, as I emphasized in §2.1, the scientific philosophy that Carnap takes for granted in ESO rejects the traditional rationalist idea that there are truths of reason we can know a priori.

---

[16] Yablo 1998 cites this sentence in support of his criticism of Quine's claim that for Carnap in ESO, 'There are physical objects' is trivially analytic. Yablo overlooks that in *Syntax* §38a, where this passage appears, the "physical language" Carnap mentions is what he calls a "coordinate language," not what he calls a "name language," and that, for the sake of the example, Carnap explicates "physical object" in the way I quoted in the text. As I explain in the text, there are other ways to explicate "physical object" in a coordinate language, and relative to some of them, 'There are physical objects' is analytic.

A fortiori, Carnap rejects the traditional distinction between a priori and a posteriori truths. (See Friedman 2007, especially p. III.) Carnap's explications of "analytic" are not intended to capture the traditional philosophical distinction between a priori and a posteriori truths, but to replace it piecemeal with distinctions between sentences that are settled as true (or false) by a set of rules we lay down for an inquiry, on the one hand, and those whose truth (or falsity) takes some empirical inquiry to establish, on the other.[17] The important question for Carnap is not whether the decision to adopt a language system in which 'There are physical objects' is analytic is true to the traditional idea of apriority, but whether such a decision is likely to be useful in our scientific inquiries.

Carnap's method of identifying and eschewing ontological questions therefore stands or falls with his analytic–synthetic distinction, which must be understood in context-sensitive, pragmatic terms, not as a conceptual analysis of the traditional idea of apriority.[18] To accept and apply Carnap's method it is enough to adopt language systems with the features he describes, to draw his trivial/nontrivial distinction for internal existence statements expressed in such language systems, and to accept his four observations about interpretation.

[17] Carnap's trivial/nontrivial distinction is based on, but distinct from, the analytic–synthetic distinction; it is the contrast between those statements that are trivially analytic or contradictory, on the one hand, and those that are not trivially analytic or trivially contradictory, but that require substantive proof or empirical inquiry to establish.

[18] A fully satisfying reading of ESO must therefore include convincing replies to Quine's trenchant criticisms (in Quine 1953b) of Carnapian transformation rules, especially Carnapian semantical rules. I have tried to develop such replies elsewhere (Ebbs 1997, chapters 4–5; Chapter 1 of this volume), with mixed success. For reasons I explain in some of the other essays in this volume, on balance I find Quine's criticisms convincing.

# PART II

## Carnap and Quine

CHAPTER 3

# Carnap and Quine on Truth by Convention

According to the standard story (a) W. V. Quine's criticisms of the idea that logic is true by convention are directed against, and completely undermine, Rudolf Carnap's idea that the logical truths of a language $L$ are the sentences of $L$ that are true-in-$L$ solely in virtue of the linguistic conventions for $L$, and (b) Quine himself had no interest in or use for any notion of truth by convention. This chapter argues that (a) and (b) are both false. Carnap did not endorse any truth-by-convention theses that are undermined by Quine's technical observations. Quine knew this. Quine's criticisms of the thesis that logic is true by convention are not directed against a truth-by-convention thesis that Carnap actually held, but are part of Quine's own project of articulating the consequences of his scientific naturalism. Quine found that logic is not true by convention in any naturalistically acceptable sense. But he also observed that in set theory and other highly abstract parts of science we sometimes deliberately adopt postulates with no justification other than that they are elegant and convenient. For Quine such postulations constitute a naturalistically acceptable and fallible sort of truth by convention. It is only when an act of adopting a postulate is not indispensible to natural science that Quine sees it as affording truth by convention "unalloyed." A naturalist who accepts Quine's notion of truth by convention is therefore not limited (as naturalists are often thought to be) to accepting only those postulates that she regards as indispensible to natural science.

## 3.1 The Standard Story of the Quine–Carnap Debate

Authors of encyclopedia entries and survey articles and books have over the years converged on a concise standard story of W. V. Quine's debate with Rudolf Carnap about the relationship between linguistic convention and logical truth. The story, which has been told and retold in countless journal articles in philosophy of language, logic, and mathematics, is that

Quine won the debate mainly, if not only, because he completely discredited the idea that logic, or any other part of science, is true by convention. As Thomas Baldwin tells the story, for instance:

> Carnap, like other logical empiricists, held that the adoption of a system of logic is fundamentally a matter of linguistic convention, so that logical truth is 'truth by convention'... Quine observed, however that the logical implications of a logical truth cannot themselves be a matter of linguistic convention, on pain of requiring an infinite number of such conventions, and thus that the role of convention in logic can amount at most to the adoption of certain fundamental principles. (Baldwin 2006, p. 77)

Paul Boghossian, Tyler Burge, Gilbert Harman, George Romanos, and Scott Soames, to name just a few, all tell roughly the same story.[1]

For Carnap, a linguistic convention is a specification of syntactical or semantical rules for a language, and a language system is a language organized according to explicitly formulated rules (Carnap 1939, pp. 6–21; Carnap 1952a, p. 432). This much, we may suppose, the standard story gets right. The problematic parts of the story concern both Carnap's understanding of the relationship between the linguistic conventions of a given language system and the logical truths of that language system, on the one hand, and Quine's understanding of and attitude toward what he calls truth by convention, on the other. These parts of the story begin with the assumption that according to Carnap,

(1) The logical truths of a language system *LS* are all and only those sentences of *LS* that are true-in-*LS* *solely in virtue of* the linguistic conventions for *LS*.

On any plausible interpretation of 'solely in virtue of', as it occurs in (1), the story goes, (1) is true *only if*

(2) The conventions for a language system *LS* *by themselves* – that is, without presupposing any additional inference rules or logical truths in a

---

[1] See Boghossian 1996, pp. 363–366, Burge 1986, pp. 699–700, Harman 1996, pp. 392–396, Romanos 1983, p. 62, and Soames 2003, pp. 264–270. (Benacerraf 1973, p. 676, tells the same basic story, though without explicitly attributing conventionalism to Carnap.) In contrast, a number of recent writings present interpretations of Carnap according to which he is not committed to the views attributed to him by the standard story. See, for instance, Creath 2007, Ben-Menahem 2006, ch. 5, Ebbs 1997, ch. 4, Friedman 2007a and 2007b, Goldfarb and Ricketts 1992, Richardson 1997 and 2007, and Ricketts 1996. Unlike the present chapter, however, none of these writings attempts to survey and discredit all minimally plausible versions of the standard story, or squarely confronts the question of how to interpret Quine's objections to various versions of the idea that logic is true by convention in light of the fact that, as I shall argue, Quine knew that his objections did not undermine any truth-by-convention theses that Carnap actually held.

metalanguage in which we specify the conventions for *LS* – logically imply all the logical truths of *LS*.

But, the story continues, in "Truth by Convention" Quine pointed out that (2) is false. Quine himself summarizes the key consideration as follows:

> Logical truths, being infinite in number, must be given by general conventions rather than singly; and logic is needed then to begin with, in the metatheory, in order to apply the general conventions to individual cases. (Quine 1963a, pp. 391–392)

This observation – henceforth, *Quine's observation*[2] – implies that (2) is not true. But, again, it seems that (1) is true only if (2) is true. Hence, the story concludes, Quine's observation completely discredits (1), the heart of Carnap's view that logical truth is truth by convention, and thereby also implies the second main part of the standard story, according to which Quine himself could not have had any interest in or need for a notion of truth by convention in his own philosophy.

This standard story of how Quine discredited Carnap's account of logical truth is superficially plausible. Nevertheless, I shall argue, it is completely wrong. Carnap did not endorse (2) or any other truth-by-convention thesis that Quine's observation undermines. Quine knew this as well as anyone. Quine's criticism of the thesis that logic is true by convention is not directed against a truth-by-convention thesis that Carnap actually held, but is part of Quine's own project of articulating the consequences of his scientific naturalism. Quine discovered that logic is not true by convention in any scientifically explanatory sense. But he also observed that in set theory and other highly abstract parts of science:

> We find ourselves making deliberate choices and setting them forth unaccompanied by any attempt at justification other than in terms of elegance and convenience. These adoptions, called postulates, and their logical consequences (via elementary logic), are true until further notice. [In such cases] postulation can plausibly be looked on as constituting truth by convention. (Quine 1963a, pp. 393–394)

For Quine, I shall argue, such postulations constitute a scientifically explanatory sort of truth by convention – a sort of truth by convention that avoids the main problem that Quine sees with Carnap's account of logical truth, namely, that it can play no explanatory role in our scientific theories.

---

[2] As Quine himself notes (Quine 1963a, p. 391 and Quine 1936, p. 104n21), Lewis Carroll makes a similar point in his parable "What the Tortoise Said to Achilles" (Carroll 1895).

If these claims are correct, they have far-reaching consequences. First, many influential arguments in philosophy of language, logic, and mathematics, including most summaries of "lessons learned" from the Quine–Carnap debate, presuppose the standard story of the debate. All of these arguments and summaries rest on false premises, and therefore need to be reassessed. Second, if the account of truth by convention that I find in Quine is successful, it solves a problem for his scientific naturalism that is often regarded as decisive. The problem is that according to the best known naturalistic strategy for justifying mathematics, the so-called indispensability argument, mathematics is justified only insofar as we need it for the natural sciences, such as physics; but it seems that quantification over sets of cardinality greater than the continuum cannot be justified in this way, so the indispensability argument leaves much of set theory unjustified (Putnam 1971, p. 347; Feferman 1993; Maddy 1997, pp. 102–107).[3] As Quine himself puts it, there are parts of mathematics – "the higher reaches of set theory" – "that share no empirical meaning, because of never getting applied in natural science" (Quine 1992, pp. 94–95). Nevertheless, he observes, the axioms of set theory that are not applicable in natural science, and hence not indispensible to it, such as the continuum hypothesis and the axiom of choice, "can still be submitted to the considerations of simplicity, economy, and naturalness that contribute to the molding of scientific theories generally" (Quine 1992, p. 95). Again, when it comes to axioms like these, according to Quine, "We find ourselves making deliberate choices and setting them forth unaccompanied by any attempt at justification other than in terms of elegance and convenience" (Quine 1963a, pp. 393–394). These adoptions constitute a sort of truth by convention that is naturalistic in the broad sense that matters for Quine's scientific naturalism, even though the postulates are not indispensible to natural science.[4] Third, and more generally, if the account of truth by convention

---

[3] Quine sometimes writes as though he endorses the indispensability argument; see, for instance, Quine 1986d, p. 400. But I am concerned in this chapter with a different strand of Quine's thinking about mathematics, one that does not seek to motivate and justify mathematical practice solely in terms of its empirical applications. This different strand is prominent in "Carnap and Logical Truth" (Quine 1963a), on which I shall focus below. But it is also evident in *Set Theory and Its Logic* (Quine 1963b), *From Stimulus to Science* (Quine 1995), and several other of Quine's writings.

[4] Quine's account of truth for the higher reaches of set theory, while not an application of the indispensability argument, may still be regarded as naturalistic in a sense that fits with the use of 'natural' in 'natural science', if "the role that mathematics plays in our empirical knowledge is the justification for counting the system *as a whole* as part of our knowledge" (Hylton 2007, p. 80; my emphasis). To defend this reading, one would have to reconcile it with Quine's claim in *From Stimulus to Science* that "no mathematical sentence has any empirical content, nor does any set of them" (Quine 1995, p. 53). And, of course, there may be other ways to view Quine's account of truth by

that I find in Quine is successful, it solves the problem faced by a Quinean scientific naturalist who finds herself deliberately adopting postulates with no justification other than that the postulates are elegant and convenient, yet who is puzzled about how it could be reasonable for her to accept and endorse the postulates thus chosen *without* presupposing, as Carnap did, that they are analytic, or immune to revision without a change in subject. The solution is to accept that the postulates are true by convention in the thin, yet explanatory sense that Quine explicates in "Truth by Convention" and "Carnap and Logical Truth."

The chapter has two main parts. In the first main part (§§3.2–3.6) I argue that the standard story is wrong. I do not dispute that Quine's observation undermines (2). To evaluate the standard story I focus, instead, on whether Carnap held a version of (1) that is true only if (2) is true. To this end, I first present my own account of Carnap's proposal that we regard logic truths as analytic (§3.2), and then examine three truth-by-convention theses that have been attributed to Carnap (§§3.3–3.5). I discuss them in order of plausibility, from least to most, and argue that Carnap is not committed to any of them. Since I am unaware of any other minimally plausible truth-by-convention theses that have been or might be attributed to Carnap, I conclude that Carnap is not committed to (2), and hence that Quine's observation does not by itself undermine Carnap's proposal that we view logical truth as analytic. In the second main part of the chapter (§§3.7–3.10), I argue that Quine's account of linguistic convention is fundamentally different from Carnap's, and that in Quine's view, some acts of adopting sentences institute a transitory and fallible yet explanatory sort of truth by convention.

## 3.2 Carnap's Analytic–Synthetic Distinction

Carnap equates the logical truths of a linguistic system with the analytic ones, and he stresses that "the analytic–synthetic distinction can be drawn always and only with respect to a language-system, i.e., a language organized according to explicitly formulated rules, not with respect to a historically given natural language" (Carnap 1952a, p. 432). What he means is that the distinction between logical truths and other kinds of truths is *directly* defined only for what he calls language systems – languages with built-in, explicitly formulated rules. The distinction can be drawn for an

---

convention as naturalistic. For my purposes in this paper, however, we need not settle these interpretive and terminological issues.

unregimented natural language, for which no rules are explicitly specified, but only *indirectly*, relative to a translation from such a language into
a language system, and hence only "with respect to" a language system.[5]
In Carnap's view, in short, to draw the analytic–synthetic distinction for
any language one must first define it for a language system with explicitly
formulated rules.

To understand Carnap's analytic–synthetic distinction (and thereby,
also, his account of logical truth) it is crucial to understand why he was
so interested in *explicitly formulated* rules. The reason, I suggest (following
Hylton 1982 and Ricketts 1982), is that Carnap wants to provide scientists
and scientifically minded philosophers with metalinguistic tools that they
can use if they wish to avoid falling into what Carnap regarded as fruitless controversies, especially the controversies of traditional metaphysics.
Carnap's attitude toward traditional metaphysics is evident in the following passage from his "Intellectual Autobiography":

> I was depressed by the disputations in which the opponents talked at
> cross purposes; there seemed hardly any chance of mutual understanding,
> let alone of agreement, because there was not even a common criterion for
> deciding the controversy.... I came to hold the view that many theses of
> traditional metaphysics are not only useless, but even *devoid of cognitive
> content*. (Carnap 1963a, p. 45, my emphasis)

The attitude toward the controversies in traditional metaphysics that
Carnap expresses in this passage reveals what I shall call *Carnap's motivating assumption*: if investigators are to agree or disagree at all, they must
share rules for evaluating their assertions. For Carnap the main task of
philosophy (or, to be more precise, the main task of the discipline that
for Carnap *replaces* what is usually called philosophy) is to propose and
apply logical and mathematical methods that inquirers can use to specify
intersubjective rules for evaluating their assertions. The task is not to show
us how to paraphrase traditional philosophical claims in a way that would
satisfy traditional philosophers, but to provide us with tools for specifying
rules that we may choose to adopt and follow in our inquiries without any
risk of falling into the sorts of fruitless disputes that are characteristic of
traditional philosophy.

---

[5] Relative to a translation $T$ from a natural language $L$ into a language system $LS$, a sentence $s$ of $L$
is analytic (L-true) or not according as a sentence $s'$ of $LS$ that $T$ correlates with $s$ is L-true-in-$LS$ or
not. A sentence $s$ of $L$ is synthetic (neither L-true nor L-false) or not relative to $T$ according as a sentence $s'$ of $LS$ that $T$ correlates with $s$ is (neither L-true-in-$LS$ nor L-false-in-$LS$) or not. For Carnap's
views on translation, see Carnap 1937, §§61–62, Carnap 1947, pp. 233–247, and Carnap 1963c.

Carnap's methods for explicitly codifying intersubjective rules were shaped by his knowledge of Gödel's incompleteness theorems. Like Gödel, Carnap distinguished between an *object language L* and a *metalanguage* in which the rules of *L* are specified and investigated. He relied on this distinction to clarify what he regarded as Wittgenstein's insight in the *Tractatus Logico-Philosophicus* (Wittgenstein 1921) that logical truths are "based solely on their logical structure and on the meaning of their terms" (Carnap 1963a, p. 25). In *The Logical Syntax of Language*, henceforth *Syntax* (Carnap 1937), Carnap proposes that we replace the vague sentence "the truth of logical statements is based solely on their logical structure and on the meaning of their terms" with a syntactical definition of logical truth (analyticity) for sentences of an object language system *LS*, stated in a metalanguage *MLS*: a sentence *s* of a (noncontradictory) language *LS* is logically true (analytic) if and only if *s* is a special sort of *syntactical* consequence in *LS* of the empty set of sentences of *LS*.

Gödel's first incompleteness theorem implies that no set of *finitary* transformation rules – formal rules that license derivations consisting of at most a finite number of sentences – can replace our pretheoretical notion of logical consequence for any language system that contains arithmetic. Carnap knew this, and therefore proposed that we adopt *transfinitary* rules, also known as ω-rules, such as DC2 of his Language I, according to which a sentence *s* of Language I that contains a free variable *v* is a logical consequence of the infinite class of sentences of Language I that result by substituting names of positive integers for *v* in *s* (Carnap 1937, p. 38). As a result, Carnap distinguished between a *derivation*, which consists of at most a finite number of sentences of a language system *LS*, and a *consequence series*, which is "a finite series of not necessarily finite classes" of sentences of *LS* (Carnap 1937, p. 39). He defines the *logical consequence* relation for a system *LS* not in terms of derivability in *LS*, but in terms of transfinitary transformation rules, such as DC2, that in general refer to infinite classes of sentences of *LS*. Carnap's use of transfinitary syntactical rules to define the logical consequence relation for a given language system *LS* is central to his account of analyticity (logical truth) in *LS*, and crucial to a proper understanding of *Syntax*'s Theorem 60C.1 – a theorem which, as I shall argue in §3.4, discredits one version of the standard story. (For more on Carnap's adoption of transfinitary rules, see Buldt 2004, Coffa 1991, Sarkar 1992, and Chapter 1 of this volume.)

Carnap saw his methods for codifying rules as part of a new scientific discipline, which he called the logic of science. In 1934, using revolutionary rhetoric that Quine would later appropriate, Carnap boldly announced that

> our own discipline, logic or the logic of science, is in the process of cutting itself loose from philosophy and becoming a properly scientific field, where all work is done according to strict scientific methods and not by means of "higher"' or "deeper" insights. (Carnap 1934, p. 46)[6]

To back this up, he emphasized that logical syntax can be arithmetized, as Gödel showed, and is itself therefore part of mathematics, hence also part of science more generally (Carnap 1937, p. 284).

To specify rules for evaluating assertions, Carnap used the clearest terms he knew. When he wrote *Syntax* Carnap did not include a general definition of truth among the rules of a language system. As soon as he learned of Tarski's method of defining truth in terms of satisfaction, however, Carnap began including what he called semantical rules – Tarski-style truth definitions, together with what he called meaning postulates, such as "if $x$ is greater than $y$ and $y$ is greater than $z$, then $x$ is greater than $z$" – among the rules of a language system. (See, for instance, Carnap 1939, §§4–77, and Carnap 1952b.) He explicated 'analytic-in-$LS$', where $LS$ is a language system comprising syntactical formation rules, syntactical transformation rules, and semantical rules, including a Tarski-style truth definition and meaning postulates, as 'logically-true-in-$LS$', or 'L-true-in-$LS$', for short. He explains the basic idea behind all of these explications as follows: "We call a sentence of a semantical [language] system $LS$ (logically true or) L-true if it is true in such a way that the semantical rules of $LS$ suffice for establishing its truth" (Carnap 1939, p. 13, with '$LS$' in place of '$S$'). Read out of context, without any understanding of Carnap's motivating assumption or of the logical methods that he used to specify the semantical rules of a language system, this sentence, especially the phrase "the semantical rules of $LS$ suffice for establishing its truth," may seem to suggest that Carnap was committed to a claim like (2) that is undermined by Quine's observation. This is a mistake, however, as I will now try to show.

---

[6] This passage by Carnap, which was only recently translated into English, was first published in German in 1934. In his 1969c "Reply to Chomsky," Quine (consciously or not) appropriates the same rhetoric. He writes that "there is no legitimate first philosophy, higher or firmer than physics, to which to appeal over the physicists' heads" (Quine 1969c, p. 303). I discuss Quine's appropriation of Carnap's rhetoric, and its relationship to Quine's criticisms of Carnap's account of logic, below.

### 3.3 Thesis 1: Meaning Determines Truth for Both Sentences and Propositions

I know of only three truth-by-convention theses that have been equated with (1) and attributed to Carnap. Consider, first, the strong metaphysical truth-by-convention thesis that Paul Boghossian attributes to Carnap – the thesis that the logical truths of a language system *LS* are all and only those sentences of *LS* that are true-in-*LS* solely in virtue of the linguistic conventions for *LS*, where a sentence *s* of language system *LS* is true solely in virtue of the linguistic conventions for *LS* if and only if for some proposition *p*, "our meaning *p* by *s* *makes it the case that p*" (Boghossian 1996, p. 365). If it is to bear on our evaluation of Carnap's project, the sentence "our meaning *p* by *s* makes it the case that *p*" must first be paraphrased in terms of our conventional choice of syntactical and semantical rules for a language system of which *s* is a part. As I noted above, in the mid-1930s Carnap adopted Tarski's method of defining truth; to specify the semantical rules for a language system *LS* he would construct a Tarski-style truth definition for *LS* and typically also lay down some meaning postulates for *LS*. According to Carnap a Tarski-style truth definition for *LS* "determine[s] for every sentence of *LS* what it asserts – in usual terms, its 'meaning'" (Carnap 1939, p. 10, with 'LS' in place of 'B-S'). I therefore propose that we paraphrase the thesis that Boghossian attributes to Carnap as follows:

> Thesis 1 (Carnapian paraphrase): The logical truths of a language system *LS* are all and only those sentences of *LS* that are *true-in-LS* solely in virtue of the linguistic conventions for *LS*, where a sentence *s* of language system *LS* is true solely in virtue of the linguistic conventions for *LS* if and only if for some proposition *p*, the linguistic conventions for *LS* by themselves – that is, without presupposing any additional inference rules or logical truths in a metalanguage in which we specify the conventions for *LS* – (a) make it the case that *s* is true if and only if *p*, and (b) thereby make *p* true.[7]

Quine's observation implies that clause (a) of Thesis 1 is false. Hence the standard story would be correct if Carnap were committed to the Thesis 1.

---

[7] Thesis 1 quantifies over propositions. Unlike Quine, Carnap did not have any principled reasons for refusing to quantify over propositions. But he preferred, when possible, to define them as equivalence classes of sentences (Carnap 1947, §§6, 33, and 40). To define propositions as equivalence classes of sentences, however, one needs to lay down semantical rules for those sentences. To save words, I shall not do this here, but shall simply assume that we need not specify semantical rules for *LS* to make sense, at least provisionally, of Thesis 1.

Let us assume, provisionally, that Carnap is committed to clause (a) of Thesis 1. (I will reject this assumption in the next section.) The question I would like to focus on now is whether Carnap is committed to there being sentences that are true by convention in the sense stated by clause (b) of Thesis 1.

Note first how implausible it is to attribute such a commitment to Carnap, who aimed above all to help philosophers *avoid* making metaphysical claims. As Boghossian himself emphasizes, it is not clear that we can even make sense of the claim that "our meaning *p* by *s* makes it the case that *p*." He asks, rhetorically:

> How can we make sense of the idea that something is made true by our meaning something by a sentence? ... are we really to suppose that, prior to our stipulating a meaning for the sentence
>
> Either Snow is white or it isn't
>
> *it wasn't the case that either snow was white or it wasn't?* Isn't it overwhelmingly obvious that this claim was true *before* such an act of meaning, and that it would've been true even if no one had thought about it, or chosen it to be expressed by one of our sentences? (Boghossian 1996, p. 365)

Boghossian writes as if he understands Thesis 1, but thinks it is incredible.[8] In my view, however, the problem with Thesis 1 is more serious: we can't give any sense at all to clause (b) of Thesis 1. Either way, if Carnap is committed to Thesis 1, he is in trouble.

Fortunately, we do not have to speculate about whether Carnap is committed to Thesis 1. He explicitly discusses the question of whether logic is true by convention in his 1939 paper "Foundations of Logic and Mathematics." His reasoning there shows that if we use his procedures to construct explications of analyticity, we will not be able to make sense of Thesis 1.

Carnap reasons as follows. We may construct a language system *LS* in one of two ways. Either (1) we first lay down a purely syntactical definition of 'L-true-in-*LS*' without making any assumptions about the meanings

---

[8] Nevertheless, Boghossian thinks that positivists, including Carnap, were committed to it. He claims that according to the positivists,

> conventional linguistic meaning, by itself, was supposed to generate necessary truths; a fortiori, conventional linguistic meaning, by itself, was supposed to generate truth. Hence the play with a metaphysical concept of analyticity [and truth by convention]. (Boghossian 1996, p. 365)

Boghossian is not alone: Alberto Coffa also attributes something like Thesis 1 to Carnap and then criticizes Carnap for accepting it (Coffa 1991, p. 312).

of the words of *LS*, and then lay down semantical rules for *LS* that fit our prior syntactical definition of 'L-true-in-*LS*', or (2) we first lay down semantical rules for some of the logical words of *LS*, and then lay down a purely syntactical definition of 'L-true-in-*LS*' that fits with those semantical rules. If (1), then to complete our construction of *LS*, we will choose semantical rules for the words of *LS* in such a way that every sentence of *LS* that is L-true-in-*LS* according to the *syntactical rules* already laid down is also true-in-*LS* according to the newly chosen *semantical rules* of *LS*. If (2), then to complete our construction of *LS*, we will adopt a syntactical definition *D* of 'L-true-in-*LS*' only if either all the sentences that are L-true-in-*LS* according to *D* are true-in-*LS* according to the semantical rules already laid down, or the semantical rules already laid down are consistent with additional semantical rules that, together with the semantical rules already laid down, imply that all the sentences that are L-true-in-*LS* according to *D* are true-in-*LS*.

Procedures (1) and (2) both enable us to construct a language system *LS* so that the sentences that are L-true-in-*LS* are true-in-*LS*. Carnap assumes that this is our goal (Carnap 1939, p. 21). Given this goal, if we follow procedure (2), we are not free to choose just *any* syntactical rules for *LS*, and so we cannot regard our syntactical definition of 'L-true-in-*LS*' as arbitrary or conventional.[9] Carnap concludes that our syntactical definition of 'L-true-in-*LS*' is "arbitrary" – just a matter of convention – only if we use procedure (1). But our syntactical definition of 'L-true-in-*LS*' does not by itself settle what the sentences of *LS* mean, because we have not yet laid down semantical rules that define truth for sentences of *LS*. Moreover, while the semantical rules that define truth for sentences of *LS* must be chosen in a way that makes some sentences of *LS* L-true-in-*LS*, and hence also true-in-*LS*, our choice of which semantical rules to adopt (hence, in effect, our decision about what sentences of *LS* are to mean) is not

---

[9] According to Carnap's Principle of Tolerance, "it is not our business to set up prohibitions, but to arrive at conventions" (Carnap 1937, p. 51), and "everyone is at liberty to build up his own logic, i.e. his own language, as he wishes. All that is required of him is that, if he wishes to discuss it, he must state his methods clearly, and give syntactical rules instead of philosophical arguments" (Carnap 1937, p. 52). Michael Potter claims that "the introduction of a theory of semantics is to be seen as a constraint on the Principle of Tolerance" (Potter 2002, p. 272; for a similar claim, see Coffa 1991, p. 321). But the reasoning of Carnap's that I just presented in the text depends on Carnap's assumption that our goal is to construct an interpretation of *LS* according to which all the sentences that are L-true-in-*LS* are true-in-*LS*. This goal is not mandatory – one might reject it. Carnap takes for granted that in practice we will almost always choose to construct language systems in which the sentences that are L-true-in-*LS* are true-in-*LS*, and that is why he writes (incautiously, I think) that if we use procedure (2), "we are indeed bound by the choice of rules in all essential respects" (Carnap 1939, p. 28).

arbitrary – it is constrained by our prior choice of the syntactical rules of
*LS* together with our prior, independent beliefs about which sentences in
our metalanguage express logical truths, or, in other words, about which
propositions are true. The key point is that if we use procedure (1) we
always construct the semantical rules of *LS* so that sentences that are
L-true-in-*LS* according to our prior syntactical definition express propositions that we *already* accept as true *before* we formulate the semantical
rules for *LS*. We are therefore unable to give any sense to clause (b) of
Thesis 1, according to which our choice of semantical rules for a language
system *LS* "makes it the case" that the propositions expressed by sentences
of *LS* that are L-true-in-*LS*, are true.

Quine presented essentially the same reasoning, applied to inter-
preted and uninterpreted geometries, in §IV of "Carnap and Logical
Truth" (Quine 1963a), written in 1954, fifteen years after Carnap's paper
"Foundations of Logic and Mathematics" was published. Quine assumed,
I suppose, that Carnap would recognize his own reasoning. Yet §IV of
"Carnap and Logical Truth" is part of Quine's criticism of the linguistic
doctrine of logical truth, which the standard story attributes to Carnap.
Hence Boghossian's version of the standard story implies, in effect, that in
"Carnap and Logical Truth," Quine uses Carnap's own arguments to crit-
icize Carnap for holding a thesis that Carnap explicitly discredited for the
same reasons in a readily available publication fifteen years before.[10]

### 3.4  Thesis 2: Meaning Determines Truth for Sentences, Not for Propositions

Carnap's reasoning in "Foundations" may nevertheless seem to leave it
open that we can specify rules for a language system *LS* that by themselves – that is, without presupposing any additional inference rules
or logical truths in a metalanguage in which we specify the conventions for *LS* – "make it the case" that certain sentences of *LS* are true
in *LS*. Gilbert Harman apparently attributes this weaker thesis, which

---

[10] In *Philosophy of Logic*, first published in 1970, sixteen years after "Carnap and Logical Truth" was
written, Quine attributes to Carnap the view that "it is language that makes logical truths true"
(Quine 1970, p. 95). Quine makes this attribution in the context of presenting his own view, and
without any detailed discussion of Carnap's account of logical truth. As we shall see below, from
Quine's naturalistic point of view, there are reasons to describe Carnap's account of logical truth in
this way. But Quine knew better than to attribute this sort of view to Carnap, even if he occasion-
ally let his rhetoric trump accuracy.

he calls "linguistic conventionalism," to the positivists, presumably including Carnap. According to Harman's understanding of linguistic conventionalism,

> In the first instance my intention makes it the case that $s$ is true and in the second place that fact about my intention (is part of what) makes it the case that $s$ means that $p$, where it is the case that $p$. This view has no commitment whatsoever as to what makes it the case that $p$. (Harman 1996, p. 394, with the sentence variable 'S' changed to '$s$')

If this characterization of linguistic conventionalism is to bear on our evaluation of Carnap's project, the phrase "my intention makes it the case that $s$ is true" must first be paraphrased in terms of our conventional choice of rules for a language system of which $s$ is a part. Here again we must keep in mind that according to Carnap a Tarski-style truth definition for $LS$ determines the "meaning" of every sentence of $LS$ (Carnap 1939, p. 10). In this case, also, however, we must find replacements for 'my intention' and 'makes it the case' and '$s$ is true'. I propose that in place of 'my intention' we use 'the conventions laid down for $LS$', where $LS$ is assumed to be some language system I have adopted, or plan to adopt. To begin with, also, I propose that in place of '$s$ is true' we use '$s$ is true-in-$LS$', and that we paraphrase 'makes it the case' as follows: 'the linguistic conventions for $LS$ by themselves – that is, without presupposing any additional inference rules or logical truths in a metalanguage in which we specify the conventions for $LS$ – logically imply that $s$ is true-in-$LS$ if and only if $p$, where $p$ is true'. These replacements yield the following Carnapian paraphrase of Harman's version of conventionalism:

> Thesis 2 (first Carnapian paraphrase): The logical truths of a language system $LS$ are all and only those sentences of $LS$ that are true-in-$LS$ solely in virtue of the linguistic conventions for $LS$, where a sentence $s$ of language system $LS$ is true solely in virtue of the linguistic conventions for $LS$ if and only if for some proposition $p$, (a) $p$ is true, and (b) the linguistic conventions for $LS$ by themselves – that is, without presupposing any additional inference rules or logical truths in a metalanguage in which we specify the conventions for $LS$ – logically imply that $s$ is true-in-$LS$ if and only if $p$.[11]

---

[11] Thesis 2 presupposes that we can quantify over the relevant propositions and suppose that they are true *prior to and independent of* our construction of $LS$. As I observed in note 7, however, Carnap preferred, when possible, to define propositions as equivalence classes of sentences. To do so, as well as to define truth for the propositions in a way that does not presuppose the syntactical or semantical rules for $LS$, we would have to ascend to a distinct metalanguage. I assume, provisionally, that

Quine's observation implies that Thesis 2 is false. Hence the standard story would be correct if Carnap were committed to Thesis 2.[12] In fact, however, Carnap was committed to rejecting Thesis 2.

To see why, consider first what Carnap would say about Thesis 2 in the period from 1932 to 1934, when he was writing *Syntax*, just a few years before Quine wrote "Truth by Convention." In this period Carnap did not yet know of Tarski's method of defining true-in-*LS*. For the logical and mathematical sentences of a language system *LS*, however, Carnap's transfinite syntactical definitions of 'analytic-in-*LS*' in *Syntax* are, in effect, syntactical surrogates for 'true-in-*LS*'.[13] To evaluate Thesis 2 in the terms available to Carnap when he wrote *Syntax*, we therefore have to use one of Carnap's syntactical definitions of 'analytic-in-*LS*' in place of 'true-in-*LS*'. The result is:

> Thesis 2 (second Carnapian paraphrase): The logical truths of a language system *LS* are all and only those sentences of *LS* that are true-in-*LS* solely in virtue of the linguistic conventions for *LS*, where a sentence *s* of language system *LS* is true solely in virtue of the linguistic conventions for *LS* if and only if for some proposition *p*, (a) *p* is true, and (b) the linguistic conventions for *LS* by themselves – that is, without presupposing any

for our purposes here we can get by without doing that, although it is not obvious that Carnap would or should accept this assumption.

[12] It is not clear from the passage by Harman that I quote in the text that he meant to commit linguistic conventionalism to clause (b) of Thesis 2. Suppose, instead, that in place of (b) in Thesis 2, we have

> (b') the conventions laid down for *LS* logically imply that *s* is true-in-*LS* if and only if *p*.

The resulting thesis, call it Thesis 2', does not seem to capture Harman's idea that my intention "determines" that *s* is true. I don't put much weight on our ideas about how to paraphrase the word 'determines', however, since that word has been used with many different meanings. For me the decisive consideration is that Thesis 2' is not vulnerable to Quine's observation, and does not even appear to give any content to the claim, central to the standard story, that logic is conventional. Moreover, few philosophers who accept the standard story are aware that (as I shall argue in the next paragraph of the main text) when Carnap wrote *Syntax*, hence before Quine first made his observation in Quine 1936, Carnap was already committed to rejecting Thesis 2, insofar as he could paraphrase it at all – that is, when the occurrence of 'true-in-*LS*' in Thesis 2 is replaced by a transfinitary syntactical definition of 'analytic-in-*LS*' of the sort that Carnap gives in *Syntax*. I therefore find it more instructive to paraphrase Harman's characterization of linguistic conventionalism by Thesis 2 than to paraphrase it by Thesis 2', or by any other thesis that is not undermined by Quine's observation.

[13] As Carnap observes, "In relation to [logical languages], certainly, 'true' and 'false' coincide with 'analytic' and 'contradictory', respectively, and are thus syntactic terms" (Carnap 1937, p. 216). The coincidence for logical languages of the *Syntax* definitions of 'analytic' and 'contractictory' with 'true' and 'false', respectively, is secured in part by Carnap's use in *Syntax* of transfinitary syntactical rules. See the penultimate paragraph of §3.2 above. Moreover, as Quine notes in "Carnap and Logical Truth," "Carnap's formulation [in *Syntax*] of logical truth proceeded along the lines ... of Tarski's technique of truth-definition" (Quine 1963a, p. 400). I will say more about this below.

additional inference rules or logical truths in a metalanguage in which we specify the conventions for *LS* – logically imply that *s* is analytic-in-*LS* if and only if *p*.

Thus rewritten, Thesis 2 is undermined by Carnap's Theorem 60C.1, which amounts to a syntactical version of Tarski's Undefinability Theorem, applied to Carnap's syntactical definition of 'analytic-in-*LS*'. Theorem 60C.1 states that *if language system LS is consistent and rich enough to express elementary arithmetic, then 'analytic-in-LS' is indefinable in LS.*[14] As Carnap himself points out, an immediate consequence of Theorem 60C.1 is that 'if the syntax of a language $LS_1$ is to contain the term 'analytic (in $LS_1$)' then it must … be formulated in a language $LS_2$ which is richer in modes of expression than $LS_1$' (Carnap 1937, p. 219, with '$LS_1$' and '$LS_2$' in place of '$S_1$' and '$S_2$', respectively). From this immediate consequence of Theorem 60C.1, it follows that the second Carnapian paraphrase of Thesis 2 (as well as a similar paraphrase of clause (a) of Thesis 1) is false: we cannot in general suppose that for any logical or mathematical sentence *s* of a language system *LS*, the linguistic conventions for *LS by themselves* – without presupposing any additional inference rules or logical truths in a metalanguage in which we specify the conventions for *LS* – logically imply that *s* is analytic-in-*LS* if and only if *p*, where *p* is true. The problem with the second Carnapian paraphrase of Thesis 2 is not that it explicitly requires that 'analytic-in-*LS*' can be formulated in *LS* – it does not – but that it implies, contrary to what we know from Theorem 60C.1, that to formulate 'analytic-in-*LS*' for a language system *LS* that is consistent and rich enough to express elementary arithmetic, we do not need to use any rules of inference or logical truths that are stronger than, and hence distinct from, those of *LS*. In short, Theorem 60C.1 undermines the second Carnapian paraphrase of Thesis 2.[15]

In his "Homage to Carnap," Quine says that he "read Carnap's *Logische Syntax* [the original German manuscript of *Syntax*] page by page as it issued from Ina Carnap's typewriter" (Quine 1971, p. 464). It is therefore difficult to believe that in 1936 Quine could have been unaware of Theorem 60C.1 of *Syntax*, or of Carnap's explicit acknowledgment of the

---

[14] For statements and proofs of Tarski's Undefinability Theorem, see Tarski 1936, pp. 253, 272–277, and Enderton 1972, pp. 228–229. Tarski's Undefinability Theorem, which is more general that Carnap's Theorem 60C.1, also implies that 'true-in-*LS*' cannot be defined in *LS*.

[15] Theorem 60C.1 also implies Quine's observation, if we suppose, as Carnap and Quine did, that the language systems of interest to us are consistent and rich enough to express elementary arithmetic.

immediate consequence of Theorem 60C.1 that I noted in the previous paragraph.[16] Yet this is exactly what we would have to believe if we were to read Quine in "Truth by Convention" as attributing the second paraphrase of Thesis 2 to Carnap and trying to discredit Carnap's account of logical truth by explaining how Quine's observation undermines Thesis 2.

As soon as Carnap learned of Tarski's method of defining truth, he would have been able to accept the first Carnapian paraphrase of Thesis 2. By then, however, or very soon thereafter, he also knew of Tarski's Undefinability Theorem, which implies that if a language system $LS$ is consistent and rich enough to express elementary arithmetic, then 'true-in-$LS$' cannot be defined in $LS$.[17] And an immediate consequence of Tarski's Undefinability Theorem is that to specify the rules that define truth for such a language system, one must use a stronger metalanguage. Hence Tarski's Undefinability Theorem undermines the first Carnapian paraphrase of Thesis 2 for the same sorts of reasons that Theorem 60C.1 undermines the second Carnapian paraphrase of Thesis 2.

It is likely that Quine learned of Tarski's Undefinability Theorem in 1936, when Tarski's paper "The Concept of Truth in Formalized Languages" was first published in German,[18] and it is virtually certain that

---

[16] Theorem 60C.1 of *Syntax* did not appear in the 1934 German edition of the book. Carnap says in the preface to the English edition of *Syntax* that Section 60 was "included in the manuscript of the German original when it was sent for publication (in December 1933) but had to be taken out for lack of space" (Carnap 1937, p. xi). Quine met Carnap in December 1932, and worked with him throughout 1933, when he was just completing *Syntax*. Quine also says, in note 12 on page 400 of Quine 1963a, that the sections of *Syntax* that were omitted from in the original German edition of 1934, including section 60, "had appeared as articles" around the same time. It is therefore difficult to believe that Quine could have been unaware of Theorem 60C.1 of *Syntax*. In any case, Carnap's Theorem 36.7, which amounts to an application of Gödel's second incompleteness theorem, was included in the 1934 German edition. Theorem 36.7 is stronger that Theorem 60C.1 and also implies that Thesis 2 is false for object language systems rich enough to express elementary arithmetic. In addition, in his 1934 lectures on Carnap, Quine notes that because of "a recent technical discovery by Gödel in foundations of mathematics," Carnap distinguishes between 'consequence of' and 'deducible from' (Creath 1990, pp. 77–78). One could not grasp this distinction, and recognize its significance for Carnap's project, without seeing that Carnap is not committed to Thesis 2.

[17] Carnap 1942 notes that Tarski analyses the liar paradox and offers us ways of avoiding it; Carnap suggests that the reader compare Tarski's treatment of the liar paradox with his own treatment of it in *Syntax*, section 60 – the section in which Carnap proves Theorem 60C.1 (Carnap 1942, p. 243). Carnap 1947 writes, "Let us look at the role of variables in an object language S. If S is given, then a metalanguage $M$ intended for the semantical analysis of S must be rich enough in relation to S. In particular, $M$ must contain variables whose ranges of values cover those of all the variables in S (and, as Tarski has shown, even go beyond this in order to make possible the definition of 'true in S')" (Carnap 1947, p. 44).

[18] Quine writes that "full details" of Tarski's method of defining truth "reached the non-Slavic world in 1936" (Quine 1963a, p. 400n12).

Quine knew the theorem by the early 1940s, when Carnap mentions it in his 1942, which Quine knew well. Moreover, when Quine learned of Tarski's theorem, he would not have missed its structural similarity to Carnap's Theorem 60C.1.[19] In "Carnap and Logical Truth," for instance, Quine observes that in *Syntax*, "Carnap's formulation of logical truth proceeded along the lines ... of Tarski's technique of truth-definition" (Quine 1963a, p. 400). Finally, Quine himself applied Tarski's Undefinability Theorem in his own logical work in the early 1950s.[20] It is therefore clear that Quine knew Tarski's Undefinability Theorem and understood its relationship to Theorem 60C.1 of *Syntax*.

In short, Carnap's proof of Theorem 60C.1 of *Syntax* and his later knowledge of Tarski's Undefinability Theorem show that Carnap was committed to rejecting Thesis 2. Quine also knew both theorems and knew that Carnap knew them. Quine therefore would not have attributed Thesis 2 to Carnap. If Carnap is committed to a truth-by-convention thesis that supports the standard story of how Quine discredits Carnap's account of logical truth, it is not Thesis 2.

### 3.5   Thesis 3: Meaning and Logic Together Show That All Logical Truths Can Be Known *a Priori*

According to Scott Soames, Carnap and other positivists hoped to explain how we can know logical truths *a priori*, without appealing to any empirical observations, by showing that all logical truths are true by convention. Soames writes,

> [The positivists'] explanation [of truth by convention] rested on two bits of linguistic knowledge that they took to be unproblematic – (i) knowledge of what we have decided our words are to mean, and (ii) knowledge that the truth of certain sentences **follows from** our decisions about what the words they contain mean. (Soames 2003, p. 264, Soames's emphasis)

[19] On p. 243 of his 1942 Carnap notes the structural similarity between his Theorem 60C.1 and Tarski's Undefinability Theorem.
[20] Quine 1952 applies Tarski's Undefinability Theorem to the language of *Mathematical Logic* (Quine 1940). Quine reports that this paper "had been figuring in my seminars for some years when, early in 1952, I was prompted by outside inquiries to write it up [for publication]" (Quine 1952, p. 141). Moreover, in Quine 1953c, a less technical work, Quine briefly sketches the consequences of Tarski's Undefinability Theorem (see Quine 1953c, p. 137). Both Quine 1952 and Quine 1953c were written before 1954, when Quine wrote "Carnap and Logical Truth," in which he repeats his objections to the linguistic doctrine of logical truth. Finally, in "Carnap and Logical Truth" itself, Quine alludes to the similarities between Carnap's Theorem 60C.1 and Tarski's Undefinability Theorem (Quine 1963a, p. 400, top, and note 12).

According to Soames, Quine's observation implies that the positivists' explanations of truth by convention face a dilemma. Either (first horn) they are *incomplete*, because, as Quine observed, explicit linguistic conventions cannot by themselves entail all the logical truths, or (second horn) they are *viciously circular*, because they simply assume that the logical generalizations and inference rules that we rely on in our metalanguage when we derive the logically true sentences of some object language system *LS* from our decisions about what the words of *LS* are to mean can *themselves* be known a priori, independent of any empirical observations.[21]

Soames emphasizes the second horn, which seems far from the standard story. But the first horn is rooted in Quine's observation. Thus the dilemma Soames presents for Carnap *includes* the standard story.

In his account of the second horn, Soames apparently assumes that the positivists, including Carnap, are committed to

> Thesis 3 (Carnapian paraphrase): The logical truths of a language system *LS* are all and only those sentences of *LS* that are true-in-*LS* solely in virtue of the linguistic conventions for *LS*, where a sentence *s* of language system *LS* is true by convention if and only if for some proposition *p*, (a) *p* is true, and (b) the conventions laid down for *LS* and the inference rules and logical truths of the metalanguage in which those conventions are specified, together imply that *s* is true if and only if *p* and thereby explain why we are entitled to accept that *s* is true on a priori grounds alone.

Against this, Soames observes that Thesis 3 presupposes, in the metalanguage, "an antecedent knowledge of logic itself." He then reasons as follows:

> Either this logical knowledge is a priori or it isn't. If it is a priori then some a priori knowledge is not explained linguistically; if it is not a priori, then our knowledge of logic isn't a priori. Either way the positivist program fails. This, in a nutshell, was one of the central arguments of Quine's paper, 'Truth by Convention.' (Soames 2003, p. 265)

---

[21] Kurt Gödel (in Gödel 1995) raises a similar but technically more sophisticated objection to Carnap's account of logical and mathematical truth. Gödel observes that "a rule about the truth of sentences can be called *syntactical* only if it is clear from its formulation, or if it somehow can be known beforehand, that it does not imply the truth or falsity of any 'factual' sentence" (Gödel 1995, p. 339). Only a *consistent* rule satisfies this requirement, since an inconsistent rule implies all sentences, including 'factual' ones. As Gödel points out, his second incompleteness theorem implies that the consistency of a language system *LS* rich enough to express elementary arithmetical truths cannot be proved using only rules and expressions that can be formulated within *LS*. He concludes from this that the truth of mathematical sentences is not settled solely by the explicit syntactical rules for any language system; the syntactical rules we lay down for a given language system *LS* cannot replace or explain our knowledge of the mathematics built into *LS*, but instead must presuppose it. For a reply on Carnap's behalf, see note 22.

This is a good objection to Thesis 3. It would therefore be a good objection to Carnap's account of logical truth if (but only if) Carnap's account of logical truth presupposes or implies Thesis 3.

The problem is that Thesis 3 makes essential use of the term 'a priori'. In *Syntax* and later works, Carnap eschews this term of traditional epistemology, along with the contrasting term, 'a posteriori' (Friedman 2007a and 2007b). To the extent that he can make sense of these terms at all, he explicates them as 'analytic (L-true) sentence' and 'true sentence that is not analytic (L-true)', respectively. As we saw earlier, however, according to Carnap, to draw the analytic–synthetic distinction for any language one must first define it explicitly for a language system with explicitly formulated rules. Hence, in Carnap's view, we have no explications of the distinction between 'a priori' and 'a posteriori' truths apart from particular language systems for which 'analytic (L-true)' is explicitly defined. The resources we rely on in a metalanguage *MLS* when we provide definitions of 'L-true in *LS*' cannot themselves be described as 'a priori' unless we have already *also* given an explicit definition of 'L-true in *MLS*'. This definition would itself have to be formulated in yet another metalanguage, *MMLS*, whose resources, in turn, could not be described as 'a priori' without some further definition in yet another metalanguage. And so on. At any given stage our definitions of a predicate 'L-true in ___', where the blank is filled by the name of a language system, presuppose our prior pragmatic assumptions about the resources we use in constructing the definition. Unless we semantically ascend to provide an explicit definition of 'L-true in ___' for our current metalanguage, say *MMLS*, we will not have not defined 'L-true in *MMLS*' (or 'a priori in *MMLS*', as Carnap explicates that phrase) for the sentences of *MMLS* that we take to be true when we use *MMLS* to define of 'L-true in *MLS*' (or 'a priori in *MLS*', as Carnap explicates that phrase) for a given language system *MLS*.

This potential regress of metalanguage definitions of the form 'L-true in ___' has been noticed by a number of commentators (Beth 1963, Goldfarb and Ricketts 1992, Ebbs 1997, and Chapter 1 of this volume). Carnap himself embraced it (Carnap 1963b). His goal was not to solve the traditional problem of how it is possible for us to have a priori knowledge, but to reject that problem, which he regarded as confused, and focus instead on specifying linguistic frameworks within which it is clear which sentences are to count as 'L-true' and which ones are to be evaluated only on the basis of empirical observations. Hence Carnap would reject Thesis 3 on the grounds that it makes use of an obscure notion of 'a priori knowledge' that he is not attempting to capture. His explications of 'analytic' as

'L-true in ___', where the blank is filled by the name of some language system, are designed not to capture its traditional philosophical meaning, but to *replace* it with definitions that we can provide in clear scientific terms, without relying on any "higher" or "deeper" explanatory or justificatory perspectives.[22]

One might object that this does not so much *answer* Soames's criticism as *avoid* it by eschewing philosophical explanation altogether. The problem, one might assume, is that any plausible interpretation of a great philosopher, including Carnap, should attribute to that philosopher some substantive *philosophical* theses. By this standard, the interpretation I just sketched, according to which Carnap is not putting forth any substantive philosophical theses about the ground of logical or mathematical truths, but simply offering us linguistic frameworks for which 'L-true' is explicitly defined, will seem implausible.

What this objection overlooks, however, is that Carnap was a revolutionary philosopher inspired by Wittgenstein's *Tractatus*, in which Wittgenstein writes that "the correct method in philosophy [is] to say nothing except what can be said, i.e. propositions of natural science – i.e. something that has nothing to do with philosophy – and then, whenever someone else wanted to say something metaphysical, to demonstrate to him that he had failed to give a meaning to certain signs in his propositions" (Wittgenstein 1921, pp. 73–74). Wittgenstein notes that this method "would not be satisfying to the other person – he would not have the feeling that we were teaching him philosophy" (Wittgenstein 1921, p. 74). I think Carnap shares Wittgenstein's view of the correct method in philosophy – so much so that Carnap never explicitly addresses traditional philosophical concerns, but, instead, recommends that we *replace* these concerns with clearer ones formulated in purely scientific terms. Recall, for instance, his bold announcement that

> our own discipline, logic or the logic of science, is in the process of cutting itself loose from philosophy and becoming a properly scientific field, where all work is done according to strict scientific methods and not by means of "higher" or "deeper" insights. (Carnap 1934, p. 46)

It is not a serious, informed criticism of the author of this credo to point out that his account of logical truth does not solve the traditional philosophical problem of how logic can be known a priori.

---

[22] Gödel's objection to Carnap, as formulated in note 21, presupposes a version of the traditional philosophical challenge to say how we come to know mathematical truths a priori, and is therefore based in a misunderstanding of Carnap's goals. For a different but compatible reply to Gödel's objection on Carnap's behalf, see Goldfarb 1995, pp. 329–330.

Quine knew that Carnap did not aim to solve the traditional problem of how logic can be known a priori. In his 1934 lectures on Carnap,
speaking as a disciple of Carnap,[23] Quine rejects the idea that an analytic
sentence is a priori because it has some special "inward necessity." He proposes that we think about the relation between a priority and analyticity
in a more pragmatic, less metaphysical way, by starting with sentences
that we want to *treat* in a given context of inquiry as settled independent of particular observations, and hence as 'a priori', in a thin, practical,
purely *methodological* sense of that word, and then constructing definitions that yield those sentences as deductive consequences (Quine 1934,
p. 65; see also Quine 1936, p. 102).

We may summarize these points as follows. Carnap's account of 'analytic' or 'L-true' is language-system-relative. Carnap did not aim to ana-
lyze or explain the traditional notion of an a priori truth in terms of his
notion of 'L-true'. Instead, he eschewed the traditional phrase 'a priori
truth', which he regarded as vague and confused. Guided by his motivating assumption that if investigators are to agree or disagree at all, they
must share rules for evaluating their assertions, he proposed that we *replace*
the traditional phrase 'a priori truth' with the particular phrases of the
form 'L-true in ___', where the blank is filled by a name of some language
system. Hence Carnap is committed to rejecting Thesis 3. Moreover, as
Quine's 1934 lectures on Carnap (Quine 1934) show, Quine knew that
Carnap was committed to rejecting Thesis 3. If Carnap is committed to a
truth-by-convention thesis that supports the standard story of how Quine
discredits Carnap's account of logical truth, it is not Thesis 3.

### 3.6 Conclusion of the Arguments in §§3.2–3.5

We have now seen that none of the Theses 1–3 supports the standard story
of how Quine discredits Carnap's account of logical truth. These are the
only minimally plausible truth-by-convention theses that have been attributed to Carnap by philosophers who believe that Quine's observation discredits a thesis to which Carnap is committed. Moreover, I do not know
of any other minimally plausible truth-by-convention theses that might
be attributed to Carnap. I therefore conclude that the first main part of
the standard story of the Quine–Carnap debate – the claim that Quine's

---

[23] In Quine's "Homage to Carnap," Quine says that he was a disciple of Carnap for six years, beginning in 1932 (Quine 1971, p. 464). During this period he wrote both his lectures on Carnap (Quine
1934) and "Truth by Convention" (Quine 1936).

observation is aimed at and directly discredits a truth-by-convention thesis to which Carnap is committed – is wrong.

As I stressed above, Carnap's account of logical truth is motivated by
his desire to offer scientists and other inquirers metalinguistic tools and
concepts they can use to lay down explicit linguistic rules for conducting
their inquiries, and thereby to avoid what he regarded as fruitless controversies – the sort that stem from a lack of clear, shared rules for evaluating
assertions. According to Carnap, the logically true sentences of a language
system are by definition those that follow from conventions that we institute by specifying what the rules of the language system are to be, together
with the (perhaps unformalized) logical truths and rules of inference in
the metalanguage in which we specify those rules. This account of logical
truth does not imply that our adoption of rules for a linguistic framework
somehow "make" the logical truths of a language system true, for reasons
(sketched in §3.3 above) that Carnap himself explained in "Foundations of
Logic and Mathematics" (Carnap 1939). Nor is it undermined by Quine's
observation, as Carnap's Theorem 60C.1 of *Syntax* and his knowledge of
Tarski's Undefinability Theorem show (see §3.4 above). The point of the
account is not to explain what "makes" logical truths true, or to show that
our knowledge of logical truths can be based on linguistic conventions
alone, without relying on logical truths and rules of inference in a metalanguage in which we specify those rules, but to provide inquirers with
metalinguistic resources they can use to articulate language systems that
facilitate their inquiries.[24]

I shall not try to clarify and defend this interpretation of Carnap
in more detail here, since I want to focus now on Quine. If, as I have
argued, Quine's observation does not discredit a thesis to which Carnap
is committed, and Quine himself knew this, then how should we understand Quine's interest in and criticisms of various truth-by-convention
theses?

### 3.7   Quine's Motivation for Asking Whether
### Logic Is True by Convention

"My dissent from Carnap's philosophy of logical truth," Quine writes,
"is hard to state and argue in Carnap's terms" (Quine 1963a, p. 385). The
difficulty Quine alludes to in this revealing but overlooked sentence –

---

[24] For a more complete explanation and defense of this interpretation of Carnap's account of logical
truth, see Ebbs 1997 and Chapter 1 of this volume.

the very first sentence of the Schilpp volume version of "Carnap and Logical Truth" – shows up in the way he later summarizes his dissent from Carnap's philosophy:

> I am as impressed as Carnap with the vastness of what language contributes to science and to one's whole view of the world; and in particular I grant that one's hypothesis as to what there is, e.g. as to there being universals, is at bottom just as arbitrary or pragmatic matter as one's adoption of a new brand of set theory or even a new system of bookkeeping. Carnap in turn recognizes that such decisions, however conventional, "will nevertheless usually be influenced by theoretical knowledge." But what impresses me more than it does Carnap is how well this whole attitude is suited also to the theoretical hypotheses of natural science itself, and how little basis there is for a distinction. (Quine 1963a, pp. 405–406)

Carnap responds to Quine's suggestion in this passage by once more urging his readers, including Quine, to distinguish between analytic statements, which are not scientific theories, and synthetic ones, which may be (Carnap 1963c, p. 917). This response from Carnap should not have been surprising to Quine, who knew as well as anyone that Carnap regarded his explications of analyticity as proposals, not theories that can be evaluated as true or false. And Quine all but anticipates the response when he stresses that his dissent from Carnap's philosophy of logical truth "is hard to state and argue in Carnap's terms" (Quine 1963a, p. 385, quoted above). Why then does Quine seem to suggest that according to Carnap, analytic statements are, or should be viewed as, theories? And if, as I have argued, Quine knows that Carnap does not believe that logic is true by explicit linguistic convention, why does Quine examine and criticize the thesis that logic is true by explicit linguistic convention?

For a while I seriously considered the possibility that Quine deliberately misled his readers by insinuating, even if not explicitly stating, that Carnap himself regarded his explication of logical truth as an explanation of logical truth on a par with the theoretical hypotheses of natural science, and hence that Carnap's explication of logical truth is vulnerable to Quine's observation, which shows that it "leaves explanation unbegun" (Quine 1963a, p. 390). This was a desperate interpretive move, but until recently I saw no good alternative to it.[25] Like most other interpreters,

---

[25] In Ebbs 1997 I interpreted Quine – wrongly, I now believe – as assuming that Carnap's definitions of analyticity are attempts to explain the truths of logic. This interpretation of Quine is widely accepted. Yemmima Ben-Menahem writes, for instance, that "there can hardly be a more serious misunderstanding of [Carnap's] position than to take the notion of truth by convention literally. But this is precisely what Quine seems to be doing [in "Truth by Convention"]" (Ben-Menahem

I assumed that Quine's naturalism is incompatible with any sort of truth by convention. And, even though I realized that Carnap did not accept any of the Theses 1–3, I still supposed it was Carnap, not Quine, who was really interested in the idea of truth by convention. Now I think I have a better interpretation, one that explains why Quine twice tried to make sense of the idea, despite Carnap's explanations of why the idea is not interesting or helpful.

On my new interpretation, Quine begins with what he regarded as the deepest motivation for all of Carnap's work – the revolutionary attitude that all perspectives independent of science must be rejected as unclear or illusory. Carnap emphasized that his own new discipline – what he called the logic of science (Carnap 1937, p. 279) – is *itself* part of science, since it consists in *pure* logical syntax, which can be arithmetized using Gödel's techiques, and *descriptive* logical syntax, which can be viewed as a branch of applied mathematics, or physics (Carnap 1937, p. 284). I see Quine as inspired by Carnap's vision of a new kind of philosophical discipline in which "all work is done according to strict scientific methods and not by means of 'higher' or 'deeper' insights" (Carnap 1934, p. 46).[26] Quine sought both to clarify this vision, and to find out how much of Carnap's work can be made consistent with it.

The unifying principle of Quine's investigation is *scientific naturalism* – "the recognition that it is within science itself, and not in some prior philosophy, that reality is to be identified and described" (Quine 1981, p. 21). To understand this characterization of scientific naturalism, it is crucial to keep in mind that, as Quine himself points out in another context, "the aim of science is not the indiscriminate amassing of truths; science is selective and seeks the truths that count for most, either in point of intrinsic interest or as instruments for coping with the world" (Quine 1982, p. 1). Quine's radical break with Carnap is rooted in Quine's view of how a scientific naturalist selects and formulates the truths that "count for most." He begins with the obvious point that the truths that "count for most" for a scientific naturalist are just the ones she affirms when, in the course of her inquiries, she adopts particular scientific theories and hypotheses.

2006, p. 234). Alan Richardson writes, more cautiously, that "insofar as Quine views such notions as 'convention' as playing a genuine philosophically explanatory role for Carnap, he is mistaken" (Richardson 1997, p. 162). These readings of Quine face a serious challenge: it is not at all plausible to think Quine could have missed that for Carnap, analytic sentences are not explanatory. Why then did Quine twice systematically investigate the proposal that logic is true by convention in an explanatory sense?

[26] As I mentioned in note 6 above, this passage by Carnap, which was only recently translated into English, was first published in German in 1934. In his 1969c "Reply to Chomsky," Quine uses the same rhetoric, but in the service of a different view of science.

All these truths may be called 'explanatory' in one good, though very broad, sense of that flexible word. The truths that are 'explanatory' in this encompassing sense are just the ones that it is the aim of science to affirm. It follows that we have no grip on what it is to be 'explanatory' in this encompassing sense apart from our own ongoing scientific inquires. In particular, if in the course of our inquiries we set forth some axioms or postulates "unaccompanied by any attempt at justification other than in terms of elegance and convenience" (Quine 1963a, p. 393), then these truths count as 'explanatory' in the same broad sense that theories and hypotheses in physics do.

In *Word and Object*, Quine sketches a general picture of scientific method that fits with this interpretation. He equates scientists' search for "simple" theories with their search for the "likeliest explanation" (Quine 1960, p. 19), and emphasizes that the *only* criterion of whether a scientific theory offers us a likely explanation is "scientific method itself, unsupported by ulterior controls" (Quine 1960, p. 23). In short, for Quine, scientific method is "the last arbiter of truth" (Quine 1960, p. 23). I see these remarks as sketches of a picture of science that was already presupposed in Quine's earliest work, including his lectures on Carnap (Quine 1934) and his paper "Truth by Convention" (Quine 1936). From Quine's standpoint, as I now understand it, the central problem with Carnap's analytic–synthetic distinction is that to classify a truth as analytic in Carnap's sense is to imply that is has no explanatory role, even in the very broad sense of 'explanatory' that I just sketched, and therefore is of no interest or significance to a Quinean scientific naturalist.[27] Yet, as we saw in the passage I quoted at the start of this section, Quine agreed with Carnap that "one's hypothesis as to what there is, e.g. as to there being universals, is at bottom

---

[27] This is how Quine should reply to Richardson's claim that "no one should feel obligated to reject the analytic–synthetic distinction simply because she wants to avoid 'first philosophy.' Quine's naturalism may not support an analytic–synthetic distinction, but opposing views, such as Carnap's, ought not be assimilated all too easily to 'first philosophy.' Certainly, Carnap took his philosophy to be every bit as scientifically respectable as Quine's. Far from trying to ground science and something more certain, Carnap sought to bring scientific status into philosophy" (Richardson 1997, p. 162). The problem is that Carnap's definition of 'analytic-in-*LS*' is designed to explicate a conception of justification for accepting statements that is independent of the statement's explanatory contribution to a scientific theory – a conception of justification that Quine associates with first philosophy. A similar problem shows up in Alexander George's interpretation of Quine's criticisms of Carnap. George argues that according to Quine, "There is no qualitative distinction between, say, disputes about whether the Earth is flat and those about whether to employ the concepts *phlogiston*, *nirvana*, or *analyticity* in one's account of reality" (George 2000, p. 19). On George's reading, Quine views Carnap's account of analyticity as part of a scientific theory that is on a par with, but less preferable than, Quine's own theory. But if in Quine's view, as I argue, Carnap's talk of 'analyticity' is explanatorily empty, then according to Quine it cannot be seen as part of an alternative scientific theory. In that respect, it is much worse off than talk of phlogiston.

just as arbitrary or pragmatic a matter as one's adoption of a new brand of set theory or even a new system of bookkeeping" (Quine 1963a, pp. 405–406). He was therefore committed to there being a scientifically legitimate sense in which even our most artificial and arbitrary acts of adopting sentences in the course of our scientific theorizing are explanatory.

### 3.8 The Thesis That Quine Aims to Undermine in "Truth by Convention"

In "Truth by Convention," Quine examines Carnap's account of logical truth from the point of view of Quine's nascent scientific naturalism. His examination presupposes

> Thesis 4: The logical truths of a language system *LS* are all and only those sentences of *LS* that are true-in-*LS* solely in virtue of the linguistic conventions for *LS*, where a sentence *s* of language system *LS* is true by convention if and only if the conventions laid down for *LS* and the inference rules and logical truths of the metalanguage in which those conventions are specified together imply that *s* is true and thereby explain why *s* is true, in some scientifically legitimate sense of 'explain'.

In "Truth by Convention" Quine does not say what would count as a scientifically legitimate sense of 'explain', but I suggest he was thinking along the lines that I sketched in the previous section. He learned from Carnap's *Syntax* that the holism of theory testing implies we may choose to affirm *any* sentence, provided we fit that choice with other choices to construct a language system that furthers our investigations (Carnap 1937, pp. 317–318). And he agreed with Carnap that a free choice to accept a sentence is different from an acknowledgment that a given sentence follows from or is supported by other sentences we already accept. But he disagreed with Carnap about how to characterize that difference. For Quine, any sentence we accept as part of one of our best current theories is 'explanatory' in the broad sense that matters for scientific naturalism, and none is immune to revision without a change in topic. To work out the details of a principled scientific naturalism, Quine therefore needed to clarify which parts of our theories are, or can be, freely chosen, and how to think about the status of truths thus adopted. This led him to develop an account of truth by linguistic convention.[28]

---

[28] Quine's account of truth by convention is part of his more general account of "the extent of man's conceptual sovereignty – the domain within which he can revise theory while saving the data" (Quine 1960, p. 5). When Quine explores "the extent of our conceptual sovereignty," he presupposes the broadly holistic picture of science that he sketches in §5 of "Two Dogmas of Empiricism"

For Quine a linguistic convention is not a Carnapian specification of explicit rules for conducting one's inquiries, but a special sort of decision to affirm (what one now takes to be) a truth in the course of one's scientific theorizing. To explain his special notion of a linguistic convention, Quine first stipulates that

> in point of *meaning* ... as distinct from connotation, a word may be said to be determined to whatever extent the truth or falsehood of its contexts is determined. Such determination of truth or falsehood may be outright, and to that extent the meaning of the word is absolutely determined; or it may be relative to the truth or falsehood of statements containing other words, and to that extent the meaning of the word is determined relatively to those other words. (Quine 1936, p. 89)

Taking these stipulations about the meaning of 'meaning' for granted, Quine observes that "a definition endows a word with complete determinacy of meaning relative to other words." Nevertheless,

> the alternative is open to us, on introducing a new word, of determining its meaning *absolutely* to whatever extent we like by specifying contexts which are to be true and contexts which are to be false. In fact, we need specify only the former: for falsehood may be regarded as a derivative property depending on the word '~', in such wise that falsehood of '—' simply means truth of '~ —'. Since all contexts of our new word are meaningless to begin with, neither true nor false, we are free to run through the list of such contexts and pick out as true such ones as we like; those selected become true by fiat, by linguistic convention. For those who would question them we always have the same answer, "You use the word differently." (Quine 1936, pp. 89–90)

In this passage Quine assumes both that the new words we introduce – words that we take to be meaningless until we decide which of the sentences in which they occur are to be true – may have the same spellings

---

(Quine 1953b). Ben-Menahem writes, "In the context of the philosophy of science, for instance in Popper's writings, the term 'conventionalism' has been used interchangeably with the underdetermination associated with Duhem's philosophy" (Ben-Menahem 2005, p. 246). Perhaps this use of 'conventionalism', together with Quine's holistic account of science, are what Barry Stroud has in mind when he writes, "[Quine's] attack on the conventionalist consists in outdoing him, in espousing 'a more thorough pragmatism' (*FLPV*, p. 46).... There is 'empirical slack' in all our beliefs, and since the whole body of our knowledge is 'under-determined' by experience, accepting or rejecting particular items will always be a matter of decision" (Stroud 1969, p. 83). This is a legitimate use of 'convention', one that it helps to keep in mind when interpreting Quine's remarks about truth by convention in "Truth by Convention" and "Carnap and Logical Truth," but it does not by itself help us to see what Quine could mean by his talk of "truth by convention unalloyed" (Quine 1963a, p. 395). For that we need the distinctions I shall highlight in §3.8 and §3.9 below. In the last paragraph of §3.9 I briefly explain the sense in which for Quine there is a conventional factor in our adoption of every sentence.

as some words already in use, and that the meanings we settle for such new words need not match the meanings of the words already in use. In specifying sentences that are to be true, our goal might not be to capture the meanings of words already in use, but just to preserve some of the functions of those words that we find useful.[29] In such cases, our decisions about which sentence are to be true are conventions, but they do not generate new truths, they simply preserve and clarify fruitful functions of old ones.

There are also cases in which we do not aim to preserve old usages, but to start new ones. In such cases, Quine writes, our decisions about which sentence are to be true institute "a second sort of convention, generating truths rather than merely transforming them," which "has long been recognized in the use of postulates" (Quine 1936, p. 88).[30] He has this second sort of convention in mind when he claims in the above passage that

> (C) We may settle the use of a word w by deciding which of the sentences in which w occurs are to be true. The sentences we select in this way "become true by fiat, by linguistic convention."

I take (C) to be fundamental to Quine's understanding of what a linguistic convention is, and how it can be explanatory, in the broad sense of 'explanatory' sketched above.

Quine's criticism of the thesis that logic is true by convention presupposes that a deliberate and explicit convention of the sort described by

---

[29] For a later statement of this attitude, see Quine 1960, section 53.

[30] Quine writes that "the function of postulates as conventions seems to have been first recognized by Gergonne" (Quine 1936, p. 88 n); he cites Gergonne's "Essai sur la théorie des definitions" (Gergonne 1818, pp. 1–35). In this work Gergonne gives an example of how 'implicit definitions' work:

> Si une phrase contient un seul mot dont la signification nous soit inconnue, l'énoncé de cette phrase pourra souvent suffire pour nous en révéler la valeur. Si, par example, on dit à quelqu'un qui connait bien les mots *triangle* et *quadrilatère*, mais qui n'a jamais entendu prononcer le mot *diagonale*, que *chacune des deux diagonals d'un quadrilatère le divise en deux triangles*, il concevra sur-le-champ ce que c'est qu'une diagonale. (Gergonne 1818, pp. 22–23).

[My translation: "If a phrase contains only one word whose meaning is unknown to us, the utterance of that phrase often suffices for revealing to us its meaning. If, for example, one says to someone who knows the words *triangle* and *quadrilateral*, but who never before heard the word *diagonal*, that *each of the two diagonals of a quadrilateral divide it into two triangles*, he will immediately grasp what a diagonal is."] Quine was also familiar with the work of E. V. Huntington, who taught in the Harvard mathematics department for decades up to and including the 1930s, when Quine wrote "Truth by Convention." Huntington was one of the most influential proponents of what Michael Scanlan (in Scanlan 1991) calls American postulate theory. In "Truth by Convention" (Quine 1936, pp. 82–88) Quine criticizes some of Huntington's work, but refers to the "postulate method" (Quine 1936, p. 83n) with approval.

(C) is explanatory.[31] The criticism is that the infinitely many truths of logic cannot all be generated solely from deliberate and explicit conventions, without presupposing logic in our metalanguage, and hence it "leaves explanation unbegun" (Quine 1963a, p. 390). The criticism is *not* that explicit and deliberate absolute conventions *cannot* be explanatory. On the contrary, Quine notes that "if the truth assignments were made one by one, rather than an infinite number at a time, the ... difficulty would disappear; truths of logic ... would simply be asserted severally by fiat, and the problem of inferring them from more general conventions would not arise" (Quine 1936, p. 105). The criticism, instead, is that Quine's observation rules out the possibility of generating the truths of logic from deliberate and explicit linguistic conventions alone, without presupposing logic in our metalanguage. Quine's observation shows that we cannot specify all the truths of logic solely in terms of deliberate and explicit linguistic conventions, and hence that they are not all true by convention and explanatory in the way that Quine thinks a sentence deliberately and explicitly adopted can be.[32] It follows that we cannot specify all the truths of logic solely in terms of deliberate and explicit linguistic conventions, and hence that they are not all true by convention, in the sense of (C).

---

[31] Many readers of "Truth by Convention" take Quine's endorsement of the idea that convention plays an important role in science to be merely provisional. Daniel Isaacson, for instance, writes that "much of ["Truth by Convention"] gives as sympathetic an account as possible to the notion of convention, but ultimately convention is rejected as a basis for truth" (Isaacson 1992, p. 109). This is a mistake, I believe, for reasons I explain in the text. Moreover, Quine's acceptance of truth by convention is not a youthful oversight. In "Two Dogmas of Empiricism," for instance, he endorses "the explicitly conventional introduction of novel notation for the purposes of sheer abbreviation" (Quine 1953b, p. 26) and, as I show in detail below, in "Carnap and Logical Truth" (Quine 1963a), written in 1954, he further develops and endorses the view of truth by convention that he sketched in "Truth by Convention" and applies it to set theory. In Quine 1963b he again endorses conventional choice as a source of truth in set theory. Even as late as 1986, Quine writes that "however inadequate as a first cause of logical truth, stipulation in the usual sense is unproblematic as a source of truth" (Quine 1986c, p. 206).

[32] In response to the objection that "we can adopt conventions through behavior, without first announcing them in words" (Quine 1936, p. 105), Quine writes, "In dropping the attributes of deliberateness and explicitness from the notion of linguistic convention we risk depriving the latter of any explanatory force and reducing it to an idle label" (Quine 1936, p. 106). His point is that even if some entrenched but implicit linguistic usage originated in an implicit optional or arbitrary "choice," the moment at which we faced that "choice" has long since passed unnoticed. If we now explicitly articulate that entrenched usage, we do not "generate new truths," but merely "transform old ones." Hence we cannot regard the truths expressed by the entrenched usage as optional and arbitrary in a sense that would *now* merit the label 'conventional'. To understand this criticism, it is crucial to keep in mind that the point of a linguistic convention in the sense of (C) above is not to make clear to oneself and others how one intends to use one's words (as Carnap's notion of a linguistic convention is supposed to do) or to solve a coordination problem (as David Lewis's account of convention in Lewis 1969 is supposed to do), but to "generate new truths."

This reasoning is the basis for a counterexample to the initially attractive idea that if we learn to accept a sentence when we learn its words, then the sentence is true by convention, in the sense of (C). In practice we cannot separate our acceptance of the truths of elementary logic from our understanding of how to use the logical words 'not', 'or', 'and', 'there is', and 'for all'. In "Carnap and Logical Truth," Quine illustrates this point by imagining that a lexicographer who wishes to translate a certain native sentence '$q$ ka bu $q$'. Quine remarks that

> if any evidence can count against a lexicographers adoption of 'and' and 'not' as translations of 'ka' and 'bu', certainly the natives acceptance of '$q$ ka bu $q$' as true counts overwhelmingly.... This is one more illustration of the inseparability of the truth of logic from the meanings of the logical vocabulary. (Quine 1963a, p. 387)

And in his 1986b "Reply to Herbert G. Bohnert," Quine writes that "there are sentences that we learn to recognize as true in the very process of learning one or another of the component words." His examples include sentences of the forms 'If $p$ then $p$', '$p$ or not $p$', 'not both $p$ and not $p$'. He emphasizes that "to have learned to use the particles 'if', 'or', 'and', and 'not' in violation of such sentences is simply not to have learned them" (Quine 1986b, pp. 94–95; see also Quine 1970, pp. 101–102). Nevertheless, the argument summarized in the previous paragraph shows that the truths of elementary logic are not true by convention, in the sense of (C).

By the criterion articulated in Thesis 4, then, elementary logic is not true by convention. Quine did not see this as a criticism of Carnap, however, since he knew that Carnap did not accept Thesis 4.[33] Quine took the trouble to investigate the consequences of Thesis 4 because he *himself* found it at least initially appealing when he wrote "Truth by Convention." He did not doubt that there are statements that are true by convention in a broad, scientifically explanatory sense. But he concluded that the elementary logical truths are not true by convention in any scientifically explanatory sense. This discovery left Quine with two fundamental questions.

---

[33] In "Carnap and Logical Truth," written in 1954, Quine repeats his argument that logic is not true by convention. On the standard story, Quine takes himself to be repeating a criticism of Carnap that Quine first made in "Truth by Convention." If, as I have argued, Quine knew that this criticism does not undermine a position that Carnap held, we need a different understanding of why he repeats the criticism in 1954. On my reading, in both "Truth by Convention" and "Carnap and Logical Truth" Quine criticizes an account of logical truth that he himself would have welcomed if it could have been made to work – an account according to which logic is true by convention in an explanatory sense that fits with his scientific naturalism.

(i) What is the explanatory contribution of those parts of what Carnap calls language systems, such as the elementary laws of logic, that are *not* true by convention in this explanatory sense?

(ii) Are some parts of our scientific theories both explanatory and true by convention?

Quine's answer to question (i) is well known. He views elementary logic as a science whose task is to transform and clarify laws and conceptual commitments that hold for all the sciences. When we explicitly formulate and adopt logical laws, we do not thereby undertake fundamentally new commitments, but clarify old ones. The explanatory force of our explicit formulations and adoptions of logical laws therefore depends on the explanatory role in all our scientific theorizing of our *prior* commitments to the truths that the explicit formulations express. Thus viewed, laws of logic are explanatory generalizations on a par with other scientific hypotheses (Quine 1960, pp. 161, 273).

According to the second main part of the standard story of the Quine–Carnap debate, Quine's answer to question (ii) is "no." This part of the standard story is also wrong, as I shall now try to show.

## 3.9 Quine's Explanatory Notion of Truth by Convention

Quine's most extended answer to question (ii) focuses on set theory. The role of set theory in our scientific theorizing is quite different from that of elementary logic. One difference is that set theory presupposes elementary logic. A related difference is that to construct a theory of sets, we need to use a sign that is not needed for elementary logic – the connective '∈' of set membership. Moreover, as Quine emphasizes, '∈' is "the one sign needed in set theory, beyond those appropriate to elementary logic" (Quine 1963a, p. 388). According to Quine, to construct a theory of sets is to regard certain sentences in which '∈' occurs as postulates or definitions, and hence as true by convention in an explanatory sense.

We begin our theorizing about sets with a motley of more or less coherent assumptions about aggregates, sets, and collections, and try to construct a single consistent set theory that disciplines and extends those parts of the motley that we find most interesting or fruitful. According to Quine, once we become aware of the set-theoretical paradoxes, we realize that each part of the motley conflicts with some other parts of it that we find at least initially appealing, and so none of the parts is obvious (Quine 1963a, p. 388). He concludes that to construct a theory of sets we have no choice but to *stipulate* that certain sentences in which '∈' occurs – the postulates and definitions of our theory of sets – are to be true.

Quine summarizes this consequence of his view of set theory in the following passage from "Carnap and Logical Truth":

> In set theory ... we find ourselves engaged in something very like convention in an ordinary non-metaphorical sense of the word. We find ourselves making deliberate choices and setting them forth unaccompanied by any attempt at justification other than in terms of elegance and convenience. These adoptions, called postulates, and their logical consequences (via elementary logic), are true until further notice. So here is a case where postulation can plausibly be looked on as constituting truth by convention. (Quine 1963a, pp. 393–394)

The postulates of set theory govern sentences that contain the word '∈' and thereby settle the meaning of '∈'. Hence postulation in set theory constitutes the sort of truth by convention that Quine cites in "Truth by Convention" in support of (C) above – a "sort of convention, generating truths rather than merely transforming them," which "has long been recognized in the use of postulates" (Quine 1936, p. 88).

Some set theorists would not agree that they set down their respective postulates for set theory by something like convention. Following Gödel 1947, they might claim to be guided by a rational perception of the true nature of sets. For Quine, however, "rational perception of the true nature of sets" can be no more than a metaphor, and, at best, a very misleading description of the phenomenology of set theoretical inquiry. His scientific naturalism requires that he find a less mysterious account of the epistemology of set theory. In Quine's view, set theorists in effect just *decide* which sentences containing '∈' to accept as postulates, and thereby decide what sets there are, according to their best current theory. This is not to deny that each particular decision between two or more candidate postulates is accompanied by rich theoretical reflections on the aptness and fruitfulness of the candidates.[34] The point is that after weighing all relevant theoretical considerations, set theorists have to decide which of the candidate postulates to adopt, even if it seems to them as if their decision is forced on them by the true nature of sets. Quine takes this point to be

[34] For any particular decision there will typically be quite a lot to say about how the investigator weighed the candidate postulates and why the postulate he decided to adopt is explanatory of the topic under investigation. It is here, at the level of the context-sensitive details of a decision to adopt a postulate in the most abstract parts of mathematics, that recent work on mathematical practice and mathematical explanation might be enlisted to supplement Quine's account. For an overview of some of the considerations and theories of explanation that may have application here, see Maddy 2007 and Mancosu 2008.

fundamental to a proper understanding of set theory, so much so that in the Introduction to his *Set Theory and Its Logic* (Quine 1963b), he writes, "A major concern in set theory is to *decide* … what open sentences to view as determining classes; or, if I may venture the realistic idiom, *what classes there are*" (Quine 1963b, p. 3, my emphasis).

To adopt by convention certain sentences that contain the word '∈' is to engage in what Quine calls *legislative postulation* (Quine 1963a, p. 394). Such postulation is not restricted to set theory. For instance, a logician who propounds "deviations from the law of excluded middle" also engages in legislative postulation, according to Quine.[35] Whether in set theory, logic, or some other discipline, at the time we introduce a legislative postulate, we do not take it merely to transform and clarify laws or other truths to which we were committed before we introduced the postulate. We take it, instead, to institute a *new* and *optional* theoretical commitment, a commitment that, as far as we now see, is unconstrained by the sorts of salient empirical considerations that guide our decisions about which empirical hypotheses to adopt, and that therefore "can plausibly be looked on as constituting truth by convention" (Quine 1963a, pp. 393–394).

Quine contrasts legislative postulation with discursive postulation – "mere selection, from a pre-existing body of truths, of certain ones for use as a basis from which to derive others, initially known or unknown" (Quine 1963a, p. 394). Unlike legislative postulation, discursive postulation results in linguistic conventions that do not generate new truths, but merely transform old ones, and therefore falls in the first category of linguistic conventions that Quine identified in "Truth by Convention" (Quine 1936, p. 88; discussed in §3.8 above). The contrast between legislative postulation and discursive postulation is paralleled by a contrast between legislative definition and discursive definition (Quine 1963a, pp. 394–395). Quine emphasizes that "it is only legislative definition, and not discursive definition nor discursive postulation, that makes a conventional contribution to the truth of sentences. *Legislative postulation*, finally,

---

[35] Quine writes, "For the deviating logician the words 'or' and 'not' are unfamiliar, or defamiliarized; and his decisions regarding truth values for their proposed contexts can then be just as genuinely a matter of deliberate convention as the creative set-theorists regarding contexts of '∈'" (Quine 1963a, p. 396). Quine assumes here that to deny the classical logician's rules for the words 'or' and 'not' is to change their meanings in such a way that the resulting logic is not in disagreement with classical logic. If the classical logician mistakenly tries to criticize the nonclassical logician, she should reply, "You use 'or' and 'not' differently." Morton 1973 challenges this consequence of Quine's view of the meaning of logical connectives, and his challenge could be generalized to apply to Quine's treatment of the meaning of '∈'. Although a full evaluation of Quine's account of truth by convention would have to address this deep issue in the theory of meaning, I shall not try to do so here.

*affords truth by convention unalloyed*" (Quine 1963a, p. 395, my emphasis). In this context, I suggest, 'unalloyed' means 'pure' – the point of the italicized passage is that legislative postulation institutes a pure form of the second sort of convention that Quine describes in "Truth by Convention," namely, the sort of convention that "generat[es] truths rather than merely transforming them" (Quine 1936, p. 88).

To "generate a truth" in Quine's sense is not somehow to "make" it true or guarantee that it is immune to revision without a change in subject. For Quine there is at most a difference in the degree to which conventional, pragmatic choice plays a role in our adoption of a legislative postulate, on the one hand, and our adoption of an empirical hypothesis, on the other.[36] To say that our adoption of a legislative postulate in set theory institutes a pure form of the second sort of truth by convention is to say that the adoption is unconstrained, as far as we can now see, by the sorts of salient empirical considerations that guide our decisions about which empirical hypotheses to adopt. Of course, the methodological role in our theories of a statement that we first introduced by an act of legislative postulation may change. For one thing, such a statement may later come to have many interconnections with other parts of our theorizing, and may therefore become deeply entrenched, even obvious to us, thereby losing its status as true by convention, even if we continue to regard it as true. For this reason, as Quine points out, "Conventionality is a passing trait, significant at the moving front of science but useless in classifying the sentences behind the lines. It is a trait of events and not sentences" (Quine 1963a, p. 395). The set of sentence-affirming events that are conventional for us as one time may therefore be different from the set of sentence-affirming events that are conventional for us as at another time.[37] And since a linguistic convention is on a par with any other part of our

[36] "I do not see how a line is to be drawn between hypotheses which confer truth by convention and hypotheses which do not, short of reckoning all hypotheses to the former category" (Quine 1963a, p. 397). Quine makes what is at root the same point in the passage about bookkeeping (from Quine 1963a, pp. 405–406) that I quoted in the first paragraph of §3.7.

[37] This consequence of Quine's view of truth by convention, together with his claim that a logician who propounds "deviations from the law of excluded middle" engages in legislative postulation, may seem to imply that it is only a contingent fact that elementary logic is not true by convention in Quine's sense. But Quine's observation, understood as restricted to classical logic, is an instance of the more general observation that *all* legislative postulation presupposes some background or other of rules of inferences and generalizations, whether classical or non-classical. And this more general observation implies that one cannot specify an infinitude of truths of *any* type by legislative postulation alone, without relying on additional rules of inference and generalizations in the language that one uses to specify that infinitude of truths.

scientific theory, according to Quine, it is in principle revisable without a change in topic, and hence not guaranteed to be true. Quine's scientific naturalism requires not that a given act of legislative postulation institute an enduring conventionality or guarantee truth, but that it be regarded, at the time of the act, by the person whose act it is, as explanatory in the broad sense I sketched above.[38]

## 3.10   Objections and Replies

The word 'explanatory' comes with a great deal of baggage, including common assumptions about its several different ordinary uses, and deeply entrenched associations with a multitude of overlapping, yet different philosophical theories of explanation. Depending on which of these assumptions or associations is strongest for you, you may object to my (and, I believe, Quine's) suggestion that there is a single, *univocal* sense of 'explanatory' in which the paradigmatic theoretical hypotheses of natural science, such as Newton's laws of motion, on the one hand, and a legislative postulate of set theory, on the other, are both explanatory. I urge you, however, not to let concerns over the proper use of the word 'explanatory' prevent you from accepting my interpretation of Quine's naturalistic account of truth by convention. If you like, you can replace 'explanatory' with 'counts for most in science' or 'scientifically significant'. Any of these phrases would do, provided they are understood to apply to the statements we accept in the course of our scientific theorizing, without relying on any "ulterior controls" (Quine 1960, p. 23), even 'controls' of the seemingly innocent sort that Carnap proposed. I use the word 'explanatory', despite its baggage, in the broad sense of 'what counts for most in science', both (1) to highlight Quine's radical claim, central to his scientific naturalism, that among those truths that "count for most" in science, there is no clear, scientifically significant boundary between those introduced by legislative postulation, on the one hand, and those introduced as hypotheses that purport to "explain" our empirical observations in more familiar, restrictive, senses of that word, on the other; and (2) to present Quine's criticisms of his version of the thesis that logic is true by

---

[38] To say that a given legislative postulate or definition is explanatory in this broad sense is not to deny that there may a great deal more to say about the precise sense in which it is explanatory. See note 34. All further details would be context-sensitive, however, and therefore cannot be part of Quine's general account of the sense in which they are explanatory.

convention – Thesis 4, above – in a way that sheds light on his use in those criticisms of the words 'explanation' and 'explanatory'.[39]

Even if you are willing to adopt these broad uses of 'explanation' and 'explanatory,' you might think that Quine's account of truth by convention does not yield anything worthy of the provocative label 'truth by convention'. In particular, you might object that since legislative postulates and definitions are not guaranteed to be true, Quine should not have called them *truths*, but *statements that we hold true*. For the same reason, you might think that Quine should not have claimed that legislative postulations or definitions yield *truth* by convention; he should have said, instead, that a given inquirer's legislative postulates and definitions are statements that she *holds-true* by convention.[40] This would be a mistake, however, according to Quine; a mistake on a par with saying that since our judgment that there are electrons is not guaranteed to be true, we should not call the statement that there are electrons a truth. Recall that for Quine it is in part by acts of legislative postulation that we decide not only which open sentences that contain the word '∈' to hold true, but also, and, thereby, *what classes there are* (Quine 1963b. p. 3). More generally, according to Quine, it is just as wrong for us to adopt a tentative attitude toward our legislative postulates and definitions as it is for us to "look down on the standpoint of [a] theory as make-believe" (Quine 1960, p. 22).

For Quine all of our scientific-explanatory practices, including mathematics, are *authoritative*, in the sense that for any of these practices, as Gideon Rosen puts it, "whenever we have reason to accept a statement given the proximate goal of the practice, we have reason to believe that it is true" (Rosen 1999, p. 471). There are, of course, other practices, such as astrology and theology, that Quine does not regard as authoritative. It may therefore seem that Quine faces what Rosen calls the Authority

---

[39] Recall, for instance, that in his criticisms of Thesis 4, Quine emphasizes that any attempt to specify all the logical truths by laying down explicit conventions "leaves explanation unbegun" (Quine 1963a, p. 390), and that the problem cannot be avoided by claiming that some conventions are implicit, not explicit, for the reason that "In dropping the attributes of deliberateness and explicitness from the notion of linguistic convention we risk depriving the latter of any explanatory force and reducing it to an idle label" (Quine 1936, p. 106).

[40] Thus one might be inclined to agree with Harman that "The epistemological force of a stipulative [legislative] definition is the same as the epistemological force of an assumption" (Harman 1996, p. 399), and to make the same claim about the epistemological force of a legislative postulate. Such a view would help us to make sense of the possibility of revising or rejecting a given legislative definition or postulate, since we readily grant that we can revise or reject an assumption. But it would also suggest – wrongly, from Quine's point of view – that we do not regard our legislative definitions and postulates as among the truths of our best current scientific theory.

Problem for Naturalized Epistemology – the problem of giving "some sort of principle for telling the authoritative practices from the rest" (Rosen 1999, p. 471). From this point of view, the fact that Quine provides no such principle may seem to discredit his scientific naturalism and, with it, his account of truth by convention.

The problem with this reasoning is that a "principle for telling the authoritative practices from the rest" would have to be offered from a perspective that is supposedly "higher" or "deeper" than any of the practices that it classifies as authoritative. Such a principle could therefore be offered only by someone who rejects Quine's scientific naturalism, according to which there is no such "higher" or "deeper" epistemological perspective on science (Quine 1969c, p. 303), and our last arbiter of truth is scientific method, where scientific method is itself just a collection of more or less similar methods used by participants in the practices that count for us as sciences, including logic, mathematics, physics, chemistry, and biology. In short, to adopt Quine's scientific naturalism is to regard the Authority Problem as illusory.

I have argued that for Quine our adoption of a legislative postulate has the same broad methodological legitimacy and significance as our adoption of an empirical hypothesis. According to Peter Hylton, however,

> for Quine ... to classify a sentence as true by convention is at most an historical speculation about the way in which it came to be accepted: it says nothing about the way in which evidence now bears on it, what justifies our accepting it, or what would justify our ceasing to accept it. Even in those cases where we can make sense of the idea of truth by convention, it refers to the way in which the sentence was introduced into our theory, not to its subsequent epistemological status. It is simply irrelevant to the present epistemological status of those sentences to which it applies; the notion thus marks no significant epistemological distinction. (Hylton 2007, p. 74)

These remarks conflict both with Quine's claim that he "[does] not see how a line is to be drawn between hypotheses which confer truth by convention and hypotheses which do not" (Quine 1963a, p. 397) and with Quine's claim that "conventionality is a passing trait, significant at the moving front of science but useless in classifying the sentences behind the lines" (Quine 1963a, p. 395). The former claim implies that for Quine the epistemological status of a legislative postulate is similar to the epistemological status of an empirical hypothesis. And the latter claim implies that to identify a statement as true by convention in Quine's unalloyed, or pure, sense is not, as Hylton claims, to speculate about how the statement once came to be accepted, but to describe the role of that statement now, in our present theory, at the moving front of science.[41] For these reasons, and the several others I present above, I cannot agree with Hylton that for Quine the notion of legislative postulation "is simply irrelevant to the present epistemological status of those sentences to which it applies."

I conclude that the standard story of Quine's debate with Carnap about the relationship between linguistic convention and logical truth is wrong. It is not Carnap, but Quine who is committed to there being an explanatory sort of truth by convention. A Quinean scientific naturalist is committed to affirming all the sentences of his best current science, including his legislative postulates and definitions, which he accepts by fiat, or linguistic convention, guided only by considerations of elegance and convenience. It is to highlight the special role in science of legislative postulates and definitions, while also rejecting a rationalist epistemology for them, that Quine calls them true by convention.

---

[41] We can describe a postulate adopted long ago as having been true by convention at that time, but only if we describe the adoption of the postulate from the point of view of the scientists who then adopted it. Such descriptions do not imply that the postulate is true by convention for us now, and hence do not imply that the postulate has any special epistemological significance for us now.

# CHAPTER 4

# Quine's Naturalistic Explication of Carnap's Logic of Science

If one studies Quine's epistemology without appreciating its deep connections to Carnap's logic of science, one can easily get the impression that unlike Carnap, Quine aims to preserve and clarify the traditional empiricist idea that our best theories of nature are justified by, or based on, our sensory evidence, and are for that reason likely to be true. Quine writes, for instance, that

> epistemology, or something like it, ... studies a natural phenomenon, viz., a physical human subject. This human subject is accorded a certain experimentally controlled input – certain patterns of irradiation in assorted frequencies, for instance – and in the fullness of time the subject delivers as output a description of the three-dimensional external world and its history. The relation between the meager input and the torrential output is a relation that we are prompted to study for somewhat the same reasons that always prompted epistemology; namely, in order to see how evidence relates to theory, and in what ways one's theory of nature transcends any available evidence. (Quine 1969b, p. 82–83)

Commenting on this and similar passages in Quine, Barry Stroud writes:

> The question Quine poses in terms of the "underdetermination" of the "torrential output" by the "meager input" makes essential use of a notion of epistemic priority. It is because the "information" we get as "input" does not uniquely determine the truth of what we assert as "output" that we must explain how we get from the one to the other. We could know everything included in our "evidence" without knowing any of the things asserted in our "theory." (Stroud 1984b, p. 252)

Donald Davidson agrees, and, to add further support for the interpretation, cites the following passage by Quine:

> We learn the language by relating its terms to the observations that elicit them ... this learning process is a matter of fact, accessible to empirical science. By exploring it, science can in effect explore the evidential relation

between science itself and its supporting observations. (Quine 1974, p. 37, cited in Davidson 1990, p. 70)

Like Stroud, Davidson takes this and other similar passages by Quine to show that Quine is committed to a traditional empiricist project (Davidson 1990, pp. 74, 76). I shall argue that despite its superficial plausibility, this interpretation of Quine is mistaken: to understand Quine's epistemology, one must see how it incorporates, yet also transforms, Carnap's principled rejection of the traditional empiricist idea that our best theories of nature are justified by, or based on, our sensory evidence, and are for that reason likely to be true.[1]

## 4.1  Carnap's Logic of Science

Carnap understood 'science' broadly, to mean "the totality of accepted sentences [including] not only the assertions of the scientists but also those of everyday life" (Carnap 1934, pp. 46–47). He was impressed, however, by the clarity and fruitfulness of sciences that employ formal and mathematical methods, especially physics, mathematics, and mathematical logic. He took these modern sciences as paradigms of rational inquiry, and was therefore depressed by traditional philosophical disputes, in which, he thought, "there [is] hardly any chance of mutual understanding, let alone of agreement, because there [is] not even a common criterion for deciding the controversy" (Carnap 1963a, p. 45, tense changed). He was also inspired by Wittgenstein's *Tractatus Logico-Philosophicus*, and sought to apply Wittgenstein's dictum that "the correct method in philosophy would really be the following: to say nothing except what can be said, i.e. propositions of natural science – and then, whenever someone else wanted to say something metaphysical, to demonstrate to him that he had failed to give a meaning to certain signs in his propositions" (Wittgenstein 1921, 6.53).

Starting in the 1930s, Carnap developed what he called the logic of science, guided by two main ideas:

---

[1] In addition to the passages in Carnap and Quine that I shall offer in support of this claim, we also have Quine's own testimony that even when his views diverged from Carnap's, Carnap set the theme. In his "Homage to Carnap," Quine writes:

> Carnap was my greatest teacher.... I was very much his disciple for six years. In later years his views went on evolving and so did mine, in divergent ways. But even where we disagreed he was still setting the theme; the line of my thought was largely determined by problems that I felt his position presented. (Quine 1971, p. 464)

(I) Everything that can be said is said by science.[2]

and

(II) If investigators are to agree or disagree at all, they must share clear, explicit criteria for evaluating their assertions.

Carnap's logic of science unites these ideas by offering inquirers mathematical terms and methods that they can use to specify clear intersubjective rules for expressing and evaluating their assertions. To specify rules for evaluating assertions Carnap used the clearest vocabulary he knew. When he first developed his logic of science, he eschewed any official talk of truth, since it cannot be defined for factual sentences in purely syntactical terms. Soon after he learned of Tarski's method of defining truth in terms of satisfaction (Tarski 1936), however, Carnap proposed that we include a Tarski-style truth definition, together with meaning postulates, among the rules of a language system (Carnap 1939, 1942, 1947, and 1952b). Carnap's acceptance of Tarski-style semantics does not mark a fundamental change in his approach to the logic of science (Ricketts 1996), however, and is not central to my account of how Quine criticizes and transforms Carnap's philosophy. To simplify my exposition, I shall therefore focus on the earlier, syntactical version of Carnap's logic of science.

A central goal of all versions of Carnap's logic of science is to clarify the roles of logic and mathematics in empirical science. On the one hand, Carnap agreed with the rationalists that the truth of mathematical statements is not contingent on the observation of facts (Carnap 1963a, p. 65). On the other hand, his view that everything that can be said is said by science led him to reject the rationalist assumption that there is a special mental faculty for perceiving truths of logic or mathematics. That assumption also strongly inclined him to accept a deflated version of the empiricist view that any "evidence" or "information" we adduce in support of our theories must be rooted in our sensory experiences.

---

[2] Carnap writes

> Everything that can be said about things is said by science, or, more specifically, by the special branch of science that deals with the corresponding domain of things. There is nothing else, nothing "higher" to be said about things than what science says about them. Rather, the object of the logic of science is science itself as an ordered complex of sentences. (Carnap 1934, p. 47)

For reasons I shall explain in the text below, the logic of science is itself part of science, according to Carnap, so we may express Carnap's motivating idea here with the sentence "Everything that can be said is said by science."

98                          Carnap and Quine

Carnap aimed to reconcile these seemingly incompatible commitments by articulating what he took to be Wittgenstein's idea that "the truths in logic and mathematics are not in need of confirmation by observations, because they do not state anything about the world of facts, [but] hold for any possible combination of facts" (Carnap 1963a, p. 65).

With this aim and his two guiding ideas in mind, in *Logical Syntax of Language* (Carnap 1937, henceforth *Syntax*), Carnap shows how to construct syntactic definitions of 'logical consequence' for what he called language systems – artificial languages with built-in syntactical rules. Each language system *LS* is characterized by its *formation rules*, which recursively specify the sentences of *LS*, and its *transformation rules*, which together settle, for every sentence *s* of *LS* and every set *R* of sentences of *LS*, whether or not *s* is a *consequence* of *R* in *LS*.

Transformation rules for a language system *LS* may include axioms of arithmetic or set theory. Gödel's first incompleteness theorem shows that no set of what Carnap calls *finite* transformation rules ("primitive sentences and rules of inference each of which refers to a finite number of premises," Carnap 1939, p. 165) can capture our pretheoretical notion of logical consequence for languages that are rich enough to express the truths of elementary arithmetic. Carnap knew this. He proposed that we define the logical consequence relation for such languages by laying down *transfinite* rules of transformation, also known as omega rules, which are defined for an infinite number of premises (Carnap 1937, §§14, 34, 43–45; Ebbs, Chapters 1 and 3 of this volume).[3] For language systems with transfinite rules of transformation, the consequence relation is not the same as the *derivability* relation, which is settled by *finite* transformation rules.

Once the formation and transformation rules of a language system are specified, one can define the key syntactical terms of Carnap's logic of science. These include the following (I will sometimes suppress "in *LS*" for readability):

> A sentence *s* (of system *LS*) is *valid* iff it is a consequence in *LS* of the empty set of sentences of *LS*; *s* is *contravalid* iff the negation of *s* is a consequence of the empty set of sentences. (Carnap 1937, pp. 173–174)

---

[3] Carnap used such transfinite rules to define logical consequence for Language I of *Syntax*. This strategy does not work for impredicative languages, however, including Language II of *Syntax*. With advice from Gödel but independently of Tarski, whose work on truth Carnap did not yet know, to define logical consequence for Language II Carnap constructed a syntactical definition of analyticity for Language II that is structurally similar to a Tarski-style definition of truth. See footnotes 17 and 20 of Chapter 3.

The *cognitive content* of sentence *s* in language system *LS* is the class of the non-valid sentences that are consequences of *s* in *LS*. (Carnap 1937, p. 175)

A sentence *s* in *LS* has (a) the *null content* if and only if its cognitive content is the empty set, and (b) the *total content* if and only if its cognitive content is the set of all the non-valid sentences of *LS*. (Carnap 1937, p. 176)

A sentence s is *determinate* if it is either valid or contravalid; *indeterminate* otherwise. (Carnap 1937, p. 174)

It follows that every valid sentence of a language system has the null content, and every contravalid sentence, the total content. Carnap accordingly explicates '*s* is factual' as '*s* is neither valid nor contravalid', or, more compactly, as '*s* is indeterminate'.

Among the determinate, hence nonfactual, sentences of a language system are its logical and mathematical sentences. To specify these, Carnap labels some of the transformation rules of a language system, such as the implication rule "*A* and $\ulcorner A \rightarrow B \urcorner$ together imply *B*," "*L*-rules." As I noted above, the *L*-rules of a language system may also include what we would ordinarily call mathematical axioms, such as the standard axioms for Peano arithmetic. A sentence *s* is *L-valid (analytic)* in language system *LS* iff the L-rules of *LS* together settle that *s* is a consequence in *LS* of the empty set of sentences of *LS*; *s* is *L-contravalid (contradictory)* iff the L-rules together settle that the negation of *s* is a consequence of the empty set of sentences of *LS* (Carnap 1937, §52).

Carnap proposes that we use the syntactical notion of L-validity *in place of* the suggestive but vague idea that the truths in logic and mathematics are not in need of confirmation by observations, because they do not state anything about the world of facts, but hold for any possible combination of facts.

A language system suitable for empirical science will also contain syntactical rules that settle what *forms* the *protocol sentences* may take (Carnap 1937, p. 317). Protocol sentences, such as 'That spot is red', 'This piece of paper is blue', are those a person may use to express observations. Since what is observable for one person might not be observable for another, the notion of a protocol sentence must be characterized relative to a person. A protocol sentence for person *P* contains at least one predicate *F*, such as 'spot', 'red', 'piece of paper', and 'blue', that is *observable* for *P*, in the sense that, "for suitable arguments, e.g., '*b*,' *P* is able under suitable circumstances to come to a decision with the help of few observations about a full sentence, say '*F(b)*', i.e., to a confirmation of either '*F(b)*' or '~*F(b)*'"

(Carnap 1936–1937, pp. 454–456, with '$P$' and '$N$' changed to '$F$' and '$P$', respectively). The syntactic specification of the forms that the protocol sentences may take will leave them indeterminate, of course, and hence factual in the sense explicated above.

In addition to the logical and mathematical sentences of a language system, on the one hand, and its protocol sentences, on the other, there may also be sentences that state laws of nature, or physical laws, such as 'If an iron thing is heated, it will expand'. Following David Hume and Karl Popper, Carnap observed that "the L-content of a law, by reason of its unrestricted universality, goes beyond the L-content of every finite class of protocol sentences.... The laws are not inferred from protocol sentences, but are selected and laid down" in accord with one's commitments regarding the existing protocol sentences (Carnap 1937, pp. 317–318). "Although our decision is based upon the observations made so far, nevertheless it is not uniquely determined by them" (Carnap 1936, p. 426). Such decisions are therefore partly "conventional" despite their "subordination to empirical control by means of the protocol sentences" (Carnap 1937, p. 320).

To make explicit the conventional element in our decisions as to which physical laws to accept, Carnap proposes that we build physical laws into a language system by stipulating P-rules, or physical rules, for the system (Carnap 1937, §51). With P-rules in place, we can define P-validity, as follows: a sentence s is *P-valid* if it is valid but not L-valid, and it is *P-contravalid* if it is *contravalid* but not *L-contravalid* (Carnap 1937, §52).

It follows from the above definitions that no valid sentence of a language system *LS* has cognitive content, whether or not the sentence is L-valid or P-valid in *LS*. Carnap is explicit about this. When he defines 'valid', he writes, "We do not use the term 'analytic' here because we wish to leave open the possibility that *LS* contains not only logical rules of transformation, ... but also physical rules such as natural laws" (Carnap 1937, p. 173). He then defines content in terms of validity, not L-validity (analyticity), and immediately notes that in his languages I and II, 'valid' coincides with 'analytic' (Carnap 1937, p. 175). If he had wanted to define content in such a way that a P-valid sentence counts as contentful, he would not have defined content in terms of validity.

According to Pierre Wagner, Moritz Schlick "argued that the laws of nature should not be confused with conventions" (Wagner 2009, p. 35). This is a natural first reaction to Carnap's notion of a P-rule. Natural or not, however, the reaction prevents us from seeing that Carnap's decision to define content in terms of validity, and not exclusively in terms of L-rules, is motivated by his goal of rejecting a range of traditional epistemological questions about the grounds for scientific claims. To appreciate this aspect of Carnap's logic of science, one needs to know a bit more about how Carnap proposes that we apply it to make and evaluate scientific claims.

The syntactical rules and terms sketched above are part of what Carnap calls *pure syntax*, which describes formation and transformation rules abstractly, without any empirical inquiry into relationships they bear to actual, natural languages. In pure syntax we may specify any such system we please and investigate its logical consequences in abstraction from any actual language. We may later stipulate that the sentence forms of a language system we have constructed in pure syntax are to be correlated with particular strings of sounds or marks that we can use to make claims. Once we have stipulated such correlations, a sentence of our physics, for instance,

> will be *tested* by deducing consequences on the basis of the transformation rules of the language, until finally sentences of the form of protocol-sentences are reached. These will then be compared with the protocol-sentences which have actually been stated and either confirmed or refuted by them. (Carnap 1937, p. 317)

Carnap was also interested in clarifying and addressing such questions as "Is the law of the constancy of the velocity of light in relativity theory a stipulation or a factual sentence?" (Carnap 1934, p. 47). To answer this question we must *translate* the sentence $s$ that relativity theorists actually use to express this law by some sentence $s'$ in a language system with explicitly formulated rules (Carnap 1937, p. 8). Such translation is a task for what Carnap calls *descriptive syntax*, within which investigators specify language systems suitable for raising *empirical* questions about the cognitive contents of sentences of actual, historically given, natural languages. The goal of such translations is to construct a language system that is "in agreement with the actual historical habits of speech" of the users of the historically given language (Carnap 1937, p. 228). Sentence $s$ is *factual* relative to a translation $T$ of relativity theory into a language system $LS$ if and only if $T$ translates $s$ by a factual sentence $s'$ of $LS$; sentence $s$ is a *stipulation*, hence nonfactual, relative to such a translation $T$, if and only if $T$ translates $s$ by an L- or P-primitive (hence valid) sentence $s'$ of $LS$. If $s$ is a stipulation in this sense then, relative to the translation, we will regard it as correct for inquirers to accept $s$ without any appeal to evidence, as expressed by their protocol sentences.

According to Carnap, to say that a given sentence *s* is *valid* relative to a given language system *LS* is to say either that

    (a)  *s* is a valid sentence of *LS* in the purely syntactical sense defined above,

or

    (b)  *s* is a sentence of a natural, historically given language that we may *translate* by a sentence *s'* of *LS* that is valid in sense (a).

Challenges to an inquirer's acceptance of a sentence *s* that is valid in either sense (a) or (b) are due to either

    (i)   an easily correctable misunderstanding or ignorance of the rules of *LS*, their consequences, or the translation into *LS* that is the basis for classifying *s* as valid,

or

    (ii)  a rejection of Carnap's principle that all that can be said is said in science, understood as codified and clarified by his logic of science.

Carnap dismisses challenges of type (ii). This has radical consequences for what Carnap calls "the question of verification," namely, "How can we find out whether a given sentence is true or false?" (Carnap 1936–1937, p. 420). Carnap's logic of science implies that judgments of truth or falsity can only be made from the standpoint of a language system, in accord with its rules. This is the point for Carnap of defining content in terms of validity, not exclusively in terms of L-validity (analyticity): if a sentence of the system is valid (whether L-valid *or* P-valid), then anyone who has chosen to use the system and understands its rules will regard the sentence as true, and no legitimate question can be raised about whether the sentence *is* true. (This is not to say that the rules of the language "make" the sentence true, or explain why it is true, however. Carnap eschews such assertions, which he regards as without cognitive content.) Indeterminate statements of a system are to be evaluated in accord with the accepted protocol sentences and the logical consequence relation for the system. Although one can question whether a given language system is the best, most fruitful one to use, such a question has no cognitive content – no true or false answer. It can only be regarded as a pragmatic question about whether it would further one's goals to adopt the system. Our choice of rules is in principle unconstrained by our prior practice, although we may choose to preserve and clarify some aspects of our prior practice. In accord with Carnap's principle of tolerance (Carnap 1937, §17), we are free to adopt any rules we like.

These radical consequences are all part of Carnap's account of what Quine calls the *doctrinal* side of epistemology (Quine 1969b, p. 71), which addresses what Carnap describes as the question of "how we get to know something, how we can find out that a given sentence is true or false" (Carnap 1936–1937, p. 420). Carnap's logic of science has equally radical consequences for what Quine calls the *conceptual* side of epistemology (Quine 1969b, p. 71), which addresses what Carnap identifies as "the question of meaning," namely, "Under what conditions does a sentence have cognitive, factual meaning?" (Carnap 1936–7, p. 420).

First, rules for language systems are specified in such a way that they can be regarded as part of science – the science of arithmetic – via Gödel numbering and recursion theory. Since mathematical practice, including number theory, allows mathematicians to propose, and derive consequences from, any definitions they like, Carnap takes the arithmetization of syntax to show that his principle of tolerance is part of science too (Carnap 1937, p. 52). In this way, he sees his logic of science as compatible with his guiding assumption that everything that can be said is said by science.

Second, the methodological relationships between statements of a language system are settled by its explicitly formulated rules, which (as we have seen) relate logical and mathematical statements of a system with laws of nature and accepted protocol statements. The rules of a language system are not meant to capture traditional, language-system-independent ideas about "logical consequence" or "evidential support," but to *clarify* and *replace* them.

Carnap's rejection of traditional epistemology is also reflected in his account of theory change. He writes that:

> if a sentence which is an L-consequence of certain P-primitive sentences contradicts a sentence which has been stated as a protocol-sentence, then some change must be made in the system. For instance, the P-rules can be altered in such a way that those particular primitive sentences are no longer valid; or the protocol-sentence can be taken as being non-valid; or again the L-rules which have been used in the deduction can also be changed. *There are no established rules for the kind of change which must be made*.... All rules are laid down with the reservation that they may be altered as soon as it is expedient to do so. This applies not only to the P-rules but also to the L-rules, including those of mathematics. (Carnap 1937, pp. 317–318, my emphasis)

Although we are committed more firmly to some rules of a system than we are to other rules of the system, "In this regard there are only differences of degree; certain rules are more difficult to renounce than others" (Carnap 1937, p. 318).

## 4.2   Quine on Truth by Convention and Analyticity

In 1933, at age twenty-five, Quine read the German manuscript of what was to become *Logical Syntax of Language* and discussed it at length with Carnap (Quine 1971, p. 464; Quine 1985, p. 98; Quine 1986a, p. 12). An entry in Carnap's diary in 1933 records Quine's early doubts about the difference between calling a statement analytic and noting that we are firmly committed to it (Quine 1991, p. 391). Quine had already begun what in retrospect we can see as a decades-long effort to make sense of those aspects of scientific practice that Carnap aimed to explicate with his logic of science.

In "Truth by Convention," first published in 1936, Quine grants that linguistic convention is integral to scientific method, but challenges Carnap's account of its methodological role. Quine's challenges are rooted in his scientific naturalism – "the recognition that it is within science itself, and not in some prior philosophy, that reality is to be identified and described" (Quine 1981, p. 21) – a doctrine that appears strikingly similar to Carnap's guiding assumption that everything that can be said is said by science. Yet Quine understands scientific naturalism in a way that led him ultimately to reject Carnap's idea that our only grip on truth or falsity is given by explicitly adopted rules for evaluating sentences.

In Quine's view, scientists seek the "likeliest explanation" of phenomena that interest them, and their only legitimate criterion for possessing such an explanation is "scientific method itself, unsupported by ulterior controls" (Quine 1960, p. 23). From Quine's standpoint, the scientific significance of any sentence must be understood in terms of its explanatory contribution to our best current theory. Quine looks for, and does not find, any explanatory significance to Carnap's classifications of much of logic and mathematics as L-valid. Quine therefore sees these classifications as having some "ulterior" motive – as not being motivated by a proper concern for the scientific significance, or explanatory contribution, of sentences.

In "Truth by Convention," Quine accepts that we may "generate new truths" by deliberately stipulating that certain sentences are to be true, as, perhaps, in the case of the law of the constancy of the speed of light. In such cases, he thinks, our deliberate stipulations that certain sentences are to be true make an explanatory contribution to our current theory, and are, for that reason, scientifically significant. His criticism of the thesis that logic is true by convention is not that deliberate and explicit conventions are not part of scientific method, and thereby lack explanatory significance, but that the infinitely many truths of logic cannot be generated solely from such conventions. We can generate the truths from such conventions if, in addition, we presuppose logic in our metalanguage, but we do not thereby distinguish the idea that logical truths are true by convention from the less mysterious observation that we are firmly committed to them.

As Quine knew, however, Carnap himself did not classify sentences of a language system as L-valid (or as valid, more generally) with the aim of showing that they make an explanatory contribution to our scientific theories. Instead, for Carnap the point of classifying a sentence as valid is to regard it as without content (in the syntactical sense defined above), hence also without any *explanatory* content, and thereby to prevent the question of why we accept it from arising at all. Properly formulated, for Carnap, such a question could only be a noncognitive, pragmatic question about which language system to adopt.

In any case, Carnap was committed for technical reasons to rejecting the goal of explaining why we accept the L-valid (analytic) sentences of a language system by appealing to its rules. As Carnap proved in *Syntax*, if *LS* is a consistent language system rich enough to express elementary arithmetic, we must use a richer metalanguage to define 'L-valid (analytic) in *LS*'. It follows from this observation that our explicit acts of adopting rules for *LS* do not *by themselves* imply the L-valid sentences of *LS*, or explain why they are all true. (See Chapters 1 and 3 in this volume for more details.)

Quine knew this as well as anyone. His criticism of the thesis that logic is true by convention is not directed against a truth-by-convention thesis that Carnap actually held, but is part of Quine's own effort to clarify the consequences of his scientific naturalism, and thereby to make sense for himself of Carnap's revolutionary assumption that everything that can be said is said by science.

Quine's criticism in "Two Dogmas of Empiricism" of Carnap's notion of analyticity is a further development of this effort. Contrary to what many commentators, including Benson Mates (1951), H. P. Grice and P. F. Strawson (1956), and Timothy Williamson (2007), have written, for Quine the central problem with Carnap's notion of analyticity is not that it is one of a family of unclear notions that cannot be defined in terms that are not also in the family, but that to classify truths as analytic in Carnap's sense is only to provide a recursive specification of truths that we already accept, not to explain why we accept them. Quine makes an exception for one-off cases of stipulation, which can, he thinks, be explanatory, in the sense that accepting the stipulations, as such, is part of our best current theory. (Even these, however, he did not regard as immune to revision without a change in subject.) He rejects Carnap's generalizations from such cases to cases in which we specify an infinitude of truths by adopting recursive rules. In the latter cases we must always *already* presuppose the general truths we recursively specify, and the specifications therefore do not explain them in the sense that matters to Quine's naturalism.

## 4.3 Quine's Pragmatism and the Doctrinal Side of Epistemology

Quine's rejection of Carnap's analytic–synthetic distinction engenders, as he says, "a shift toward pragmatism" (Quine 1953b, p. 20). A reader who is unaware of how closely Quine follows Carnap here might take Quine to be adopting a pragmatic theory of justification, according to which the convenience and usefulness of a theory, judged by our own best pragmatic standards, are reasons for thinking that it is likely to be true. In fact, however, as Quine later reports, when he spoke of a shift toward "pragmatism," he "was merely taking the word from Carnap and handing it back to him: in whatever sense the framework for science is pragmatic, so is the rest of science" (Quine 1991, p. 397). When Quine rejects the analytic–synthetic distinction, he allows that the pragmatic considerations Carnap describes as relevant to *selecting* and *revising* language systems are part of scientific method. For Quine, as noted above, we can judge truth only from the standpoint of our best current theory, via an application of "scientific method itself, unsupported by ulterior controls" (Quine 1960, p. 23).

This is not an answer to the traditional question "What justification do we have for thinking that our best theories are true?" but a rejection of it. Quine's attitude to this traditional question is similar in spirit to Carnap's, despite being stripped of any appeal to Carnapian transformation rules. "What evaporates," Quine writes, "is the transcendental question of the reality of the external world – the question of whether or in how far our science measures up to the *Ding an sich*" (Quine 1981, p. 22).

A related similarity emerges when one considers a common objection to the pragmatic holism about testing and revising our theories that Quine espouses in a notorious passage from "Two Dogmas": "Any statement can be held true come what may, if we make drastic enough adjustments elsewhere in the system.... Conversely, and by the same token, no statement is immune to revision" (Quine 1953b, p. 43). The objection is that if there are no Carnapian transformation rules, and Quine's radical claims about theory revision are correct, then "the internal structure of [a] theory, consisting in the interconnections of sentences with one another, is totally dissolved, and the theory becomes a mere featureless collection of sentences standing in no special relations to each other" (Dummett 1981, p. 597; see also Etchemendy 2008, p. 297).

This objection rests on two fundamental mistakes. First, Quine's holism about theory change is just a naturalized version of Carnap's observation that there are no established rules for revising a language system (Carnap 1937, p. 318). In practice, *both* Carnap and Quine would appeal to our best pragmatic sense in a given context of which statements we wish to retain and which ones we may revise in our efforts to arrive at the best overall theory. If the objection challenges Quine's view, it also challenges Carnap's. Second, and more seriously, however, the objection rests on a conflation of *implication* – or the logical consequence relation on a given formalization of our theoretical commitments – with what Gilbert Harman calls *inference* – our actual *adoption* of new statements that we take to be implied by other statements to which we are already committed. (Harman 1986, pages 1–5, Chapter 2.) Each in his own way, Carnap and Quine draw this distinction, and neither of them takes his accounts of implication to yield any immediate consequences for how an inquirer is to arrive at or revise a theory.

Despite these similarities in Carnap's and Quine's views of theory change, Quine's rejection of Carnap's analytic–synthetic distinction, and his corresponding incorporation of Carnap's pragmatism into scientific method itself, commit Quine to his doctrinal principle that we must judge truth from the standpoint of our best theory. This principle is incompatible with Carnap's principle of tolerance, and thereby marks a deep divergence between Quine's and Carnap's lines of thought.

## 4.4   Quine's Naturalism and the Conceptual Side of Epistemology

Quine's principle that we must judge truth from the standpoint of our best theory also disciplines and explains his naturalistic account of the conceptual side of epistemology.

To see how, recall first that for Quine there is no sense to be made of the idea that all of logic is true by convention. We have deeply entrenched commitments in logic, and sentences of logic are related in multifarious ways to observation sentences. Logic is therefore, in that indirect sense, empirical (Quine 1982, pp. 1–3). We don't have to go through any special investigation to affirm these laws, however, since we come to find them obvious when we learn them, and in practice we are guided by a principle of conservatism (Quine 1986f, chapter 7). These laws are part of our best current theory, and that is enough entitlement to go on. If pressed, however, we can cite their systematic import for our theorizing as among our pragmatic grounds for continuing to accept them.

Some parts of mathematics, by contrast, including the higher reaches of set theory, "share no empirical meaning," according to Quine, "because of never getting applied in natural science" (Quine 1992, p. 94). In these parts of mathematics, all we have to go on are considerations of elegance and convenience, including those of "rounding out" our mathematical theories (Quine 1986b, p. 400). Quine thinks it makes sense to call these parts of mathematics "true by convention," in a thin sense that does not imply that they are analytic or guaranteed to be true (Quine 1963a, p. 395; Chapter 3 of this volume).

As we saw above, Carnap rejects the rationalist assumption that there is a special mental faculty for perceiving truths of logic or mathematics. So does Quine, and for the same basic reason. According to Quine, "Science tells us that our only source of information about the external world is through the impact of light rays and molecules upon our sensory surfaces" (Quine 1975, p. 68). Some parts of mathematics are "justified" by their relationship to sensory evidence (Quine 1986d, p. 400; Putnam 1971, p. 347), while others, the parts that merely "round out" mathematics, are not (Quine 1986d, p. 400).

But what sense of "justified" is relevant here and in other parts of our theorizing, according to Quine? As I noted above, some readers, including Stroud and Davidson, take Quine to be committed to explaining how our theories are "based on" sensory "evidence," where "evidence" is construed as impacts at our nerve endings and the "based on" relation is akin to the traditional epistemological notion of evidential support. In a similar vein, Michael Friedman offers the following account of Quine's philosophy of mathematics:

> [According to Quine] we can view our total system of nature as a conjunction of set theory with various scientific theories standardly so-called – a conjunction which is tested as a whole by the deduction (in first-order logic) of various empirical consequences with this total system. The existential commitments of set theory are thus empirically justified to the same

extent, and in the same way, as are our "posits" of any other theoretical entities in natural science. (Friedman 2007a, pp. 112–113)

We know this cannot be the whole story, however, since, as we saw above, Quine believes that some parts of set theory never get applied in natural science. The central problem with this whole approach, however, is that Quine himself rejects it. In a reply to Davidson, for instance, Quine says that in his theory "the term 'evidence' gets no explication and plays no role" (Quine 1990, p. 78).

But how, then, should we understand Quine's detailed investigations of the relations between sentences and impacts at our nerve endings? The point of these investigations, I suggest, is not to explicate the notion of "evidence" or the traditional idea of "evidential support," but to provide a naturalistic replacement for the conceptual side of Carnap's logic of science.

First, in place of Carnap's *arithmetization* of his logical syntax, Quine offers a naturalistic *psychological* account of the actual relationships between our sentences and impacts at our nerve endings. This naturalization has for Quine the same significance as the arithmetization of logical syntax has for Carnap – it shows that Quine's description of science can be given from within science itself and thus fits with his scientific naturalism. "Epistemology in its new setting," Quine writes, "is contained in natural science, as a chapter of psychology" (Quine 1969b, p. 83).

Second, by noting that, from a behavioristic point of view, our sentences are related to other sentences and to sensory stimulation in complex, interdependent ways, Quine's account mirrors, and thereby clarifies, the doctrinal point that we can only judge truth from within our best current theory.

Consider, for instance, the relationship between physical laws and protocol sentences in Carnap's account of the logic of science. As we saw above, for Carnap "the construction of the physical system is not effected in accordance with fixed rules, but by means of conventions" (Carnap 1937, p. 320). Quine reveals the conventional character of a law not by describing it as a P-rule, but by noting that to *decide* to accept it *as a law* from the standpoint of our best current theory is, when described naturalistically, to acquire speech dispositions that link that sentence to other sentences, including one's observation sentences, in certain characteristic ways. The element of *convention* in our decision to accept a sentence as a law from the standpoint of our best current theory is *mirrored* in Quine's naturalized description by the fact that any particular set of speech dispositions fixes just one of many different possible ways of relating theoretical sentences to observation sentences that we accept. From a purely descriptive, psychological point of view on science, as Bredo Johnsen puts it, *"the only connections there are* between our evidence and our theories are *those we make* in accordance with scientific method" (Johnsen 2005, p. 87).

This way of describing the relation between "evidence" and "theory" may seem at odds with our standpoint as inquirers. When we are engaged in scientific inquiry, we do not think of our application of scientific method as a matter of making evidential connections, but of discovering, or recognizing, how things are. This observation is not in tension with Quine's naturalistic descriptions of how our theoretical sentences are related to our observation sentences, however, since for Quine those descriptions are always disciplined by, and solely in the service of, his doctrinal point that we can judge truth only from the standpoint of our best current theory. From this standpoint, we do not look down on the laws of our theory as "make-believe," for "we can do no better than occupy the standpoint of some theory or other, the best we can muster at the time" (Quine 1960, p. 22). Despite being underdetermined, our best current theory is our "ultimate parameter" (Quine 1969c) – scientific method, unsupported by ulterior controls, is our "last arbiter of truth" (Quine 1960, p. 23). From this doctrinal point of view, the sense in which our laws are conventional is just that when we adopt them we are guided by our best current understanding of scientific method, including the sorts of pragmatic decisions that Carnap identified as underdetermined by our sensory experience, yet integral to adopting any physical theory at all.

In short, the goal of Quine's naturalistic account of the relationship between theory and evidence is not to show that our best theories of nature are justified by our sensory evidence, but to show that we can describe science from within science itself in a way that mirrors and thereby clarifies his doctrinal principle that we can only judge truth from the standpoint of our best current theory.

# PART III

## *Quine*

# CHAPTER 5

# Quine Gets the Last Word

Today, nearly sixty years after the first publication of W. V. Quine's *Word and Object* (Quine 1960), two related criticisms of Quine's philosophy have become entrenched. One is that Quine's naturalistic epistemology is descriptive, not normative, and so fails to provide an account of epistemic justification. The other is that Quine's naturalistic account of meaning leads to the absurd conclusion that our sentences are meaningless, so we cannot use them to make assertions. I shall argue that these entrenched criticisms ultimately rest on a failure to grasp Quine's resolutely minimalist understanding of language use.

In Quine's view, language use – in particular, our use of language to theorize – is pragmatically indispensable, and although our sentences are associated with sensory stimulation by the mechanism of conditioned response, descriptions of such associations do not convey what it is to use one's sentences to make assertions. When Quine tries to convey his minimalist understanding of what it is to use one's sentences to make assertions, his task is therefore not one of theorizing about it, but of elucidating it. His strategy for elucidating it is to invite us, his readers, to start with our present understanding of our use of sentences to make assertions and then to *subtract* our doubtful or confused assumptions about what is required for such use. Subtracting these assumptions from our understanding of our use of sentences to make assertions, we get Quine's resolutely minimalist understanding of language use as the difference.

## 5.1 Truth and Disquotation

Passages in which Quine tries to elucidate his understanding of language use are scattered throughout his writings, but there is an unusually high concentration of them in *Word and Object*, §6. Here Quine emphasizes, "What reality is like is the business of scientists, in the broadest sense, painstakingly to surmise" (Quine 1960, p. 22). This is Quine's scientific naturalism: we use sentences to surmise, or theorize, but we have no grip on *what* to theorize apart from scientific method, loosely understood as what shapes our sense of what best explains the phenomena that interest us. In other words, scientific method is our last arbiter of truth (Quine 1960, p. 23). The truth of a given sentence cannot in general be defined in terms of scientific method, however. First, it is unlikely that there is a unique best total theory, each of whose sentences is true (Quine 1960, p. 23). Second, even if there *were* such a theory we could not use it to define truth for sentences of other theories, since "there is in general no sense in equating a sentence of a theory $\theta$ with a sentence $S$ given apart from $\theta$. Unless pretty firmly and directly conditioned to sensory stimulation, a sentence $S$ is meaningless except relative to its own theory; meaningless intertheoretically" (Quine 1960, p. 24). Quine concludes that truth can only be defined disquotationally:

> Where it makes sense to apply 'true' is to a sentence couched in the terms of a given theory and seen from within the theory, complete with its posited reality. Here there is no occasion to invoke even so much as the imaginary codification of scientific method. To say that the statement 'Brutus killed Caesar' is true, or that 'The atomic weight of sodium is 23' is true, is in effect simply to say that Brutus killed Caesar, or that the atomic weight of sodium is 23. (Quine 1960, p. 24)

In this passage Quine invites us to acknowledge that even if meaning is indeterminate, we can use our sentences to make assertions, including, but of course not only, assertions about when a truth predicate applies to them.

Quine's talk of applying 'true' only to a sentence couched in the terms of a given theory and seen from within the theory may seem to suggest that he is committed to "a relativistic doctrine of truth – rating the statements of each theory as true for that theory, and brooking no higher criticism" (Quine 1960, p. 24). But he resolutely rejects any such relativism about truth:

> The saving consideration is that we continue to take seriously our own particular aggregate science, our own particular world-theory or loose total fabric of quasi-theories, whatever it may be. Unlike Descartes, we own and use our beliefs of the moment, even in the midst of philosophizing, until by what is vaguely called scientific method we change them here and there for the better. Within our own total evolving doctrine, we can judge truth as earnestly and absolutely as can be; subject to correction, but that goes without saying. (Quine 1960, pp. 24–25)

It is no accident that Quine places this ringing rejection of relativism at the very end of §6, in which he previews his reasons for rejecting the standard assumptions that synonymy is an objective relation and that the norms of epistemic justification are prior to and independent of our best current scientific theories. By subtracting these assumptions from our understanding of our use of sentences to make assertions, Quine thinks, we get a clearer, more useful understanding of our use of language as the difference.

## 5.2 Entrenched Criticisms

According to Jaegwon Kim, however, Quine's subtractions leave us with nothing worthy of the label "epistemic justification." Kim argues that epistemic justification "manifestly is normative," since it concerns whether or not it is permissible, reasonable, or epistemically responsible for a person to hold a given belief (Kim 1988, p. 383). The problem, according to Kim, is that Quine's causal-dispositional account of the relationship between "theory" (the totality of sentences we are disposed to affirm) and "evidence" (impacts at our nerve endings) is descriptive, not normative, and hence tells us nothing about when it is permissible, reasonable, or epistemically responsible for a person to hold a given belief (Kim 1988, pp. 388–389).

There is a grain of truth to Kim's criticism: if we focus exclusively on Quine's causal-dispositional account of the relationship between "theory" and "evidence," then normativity drops out of the picture. But if we focus on what Quine tells us about epistemology from the standpoint of our own evolving theory, we can eke out a residual, context-dependent sort of normativity. For instance, in the passage from *Word and Object*, §6, in which Quine introduces his disquotational account of truth, he writes:

> [The] statement[s] 'Brutus killed Caesar' is true, or that 'The atomic weight of sodium is 23' is true ... are justifiable only by supplementing observation with scientific method. (Quine 1960, p. 24; cited in Johnsen, p. 83)

In the same spirit, Quine would agree with Gilbert Harman that an inference from one belief to another is epistemically justified if we accept it as part of our best current scientific theorizing (Harman 1976).

One might object that if this is all there is to say about justification, one would not be able to distinguish between *taking* oneself to be justified on a given occasion and *being* justified on that occasion, and hence one would not be able to see oneself as guided by norms of justification when one judges or asserts. This objection seems especially powerful if we focus on Quine's view that there is no sharp distinction between theoretical questions and pragmatic questions of language policy (Quine 1960, p. 271). On this view, a scientist working from within the standpoint of his own current theory may weigh pragmatic questions of language policy in deciding what to believe. For questions of language policy, however, there can be no objective, inquirer-independent standards of right or wrong. It is therefore unclear what, if anything, remains of the apparently indispensible distinction between a speaker's taking herself to be justified on a given occasion and her being justified on that occasion. But if nothing of that distinction remains, then nothing remains of our idea of epistemic standards that *guide* us in our efforts to judge or assert in accord with the truth. How, then, on Quine's view, can there so much as *be* an act of judging or asserting?

A similar challenge arises from the meaning-theoretic side. H. P. Grice and P. F. Strawson argue that Quine's naturalism about meaning leads to the absurd conclusion that our sentences are meaningless. "If talk of sentence-synonymy is meaningless," they argue, "then it seems that talk of sentences having a meaning at all must be meaningless too" (Grice and Strawson 1956, p. 146). This criticism, which is directed at Quine's arguments in "Two Dogmas of Empiricism," looks even more compelling after one reads chapter 2 of *Word and Object*, in which Quine argues in detail for his indeterminacy thesis. For if translation is objectively indeterminate, then the meanings of one's own sentences are objectively indeterminate. It apparently follows that when I try to "use" a sentence to assert, for instance, that a rabbit is not a rabbit part, I will inevitably fail, since my words have no determinate meaning.

This conclusion may seem inescapable if we read Quine's account of meaning as a holistic theory of the meanings of sentences. According to Michael Dummett, for example, "Quine ... inclines to a holistic view of language, according to which the full explanation of the meaning of any sentence will involve an explanation of the entire language" (Dummett 1993, p. 139). On this reading, "Meaning ... becomes for Quine something essentially ineffable. We cannot say what meaning our language, or any part of it, has, since any attempt to do so would only yield a further constituent of that to which meaning is to be assigned" (Dummett 1981, p. 596). If we agree that on Quine's view meanings are ineffable, it is unclear how we can accept his view and still take ourselves to use our sentences to make assertions.

We get a similar result without relying on Quine's indeterminacy thesis or his supposed holism about sentence meaning if we accept a claim that is central to Saul Kripke's presentation of Wittgenstein's remarks about meaning: our words are meaningful only if there are facts that determine how they ought to be used. As Kripke argues, it apparently follows from any dispositional theory of meaning, including Quine's, that our assertions of sentences are mere "jack-in-the-box" reactions to stimuli, and hence that nothing determines how our sentences ought to be used (Kripke 1982, p. 23). John McDowell agrees: if our utterance of a sentence is just a causal-dispositional reaction to stimuli, he reasons, then it is no more than a "mere brute meaningless sounding off" (McDowell 1984, p. 336). In short, if our words are meaningful only if there are facts that determine how they ought to be used, then Quine's dispositional account of meaning apparently leads to the absurd conclusion that our words are meaningless, and so we cannot use them to make assertions.

## 5.3 The Fallacy of Subtraction and Explication

These entrenched criticisms of Quine's philosophy rest on a failure to grasp his resolutely minimalist conception of language use. To see why, recall first that Quine aims to subtract confused and mistaken assumptions about meaning and justification from our understanding of our use of sentences to make assertions. Grice and Strawson's main inference – "If talk of sentence-synonymy is meaningless, then it seems that talk of sentences having a meaning at all must be meaningless too" (Grice and Strawson 1956, p. 146) – in effect reasserts some of the very views of meaning that Quine aims to subtract from our understanding of language use. Since Quine rejects Grice and Strawson's requirements on meaningfulness, he sees no reason to accept their conclusion that our use of sentences to make assertions is incompatible with there being no objective synonymy relation between sentences. Their main inference is an instance of what Quine calls the "fallacy of subtraction," whereby it is argued that if we start with a sentence that we use to make assertions and we subtract our assumption that sentence synonymy is an objective relation, we are left with a meaningless sentence, that is, one that we cannot use to make assertions (Quine 1960, p. 206).

Quine sees no incompatibility between his thesis that translation is indeterminate and our use of sentences to make assertions. He therefore spends very little time on the fallacy of subtraction. In *Word and Object*, he even shows some exasperation at having to deal with the fallacy at all. He had already done so, he thought, in earlier publications, especially his influential paper "On What There Is," where he wrote, "I feel no reluctance toward refusing to admit meanings, for I do not thereby deny that words and statements are meaningful" (Quine 1953a, p. 11).

The fallacy of subtraction will escape us if, like Dummett and many others, we suppose that Quine's account of meaning is a holistic theory of sentence meaning. For then we will believe that Quine is committed to the conclusion that sentence meaning is ineffable, and hence that we cannot use our sentences to make assertions. This reading of Quine is mistaken, however, as Hilary Putnam points out in "Meaning Holism" (Putnam 1986, pp. 423–424, which refers to Goldfarb 1983); like Wittgenstein's notorious suggestion that "meaning is use," Quine's account of meaning and his thesis of the indeterminacy of translation do not amount to a substantive theory of sentence meaning, but to a denial of the appropriateness of theorizing about sentence meaning. We have confirmation of Putnam's reading in Quine's approval of it (Quine 1986e), as well as in Quine's charge that Grice and Strawson commit the fallacy of subtraction.

Kripke's jack-in-the-box objection also commits a version of the fallacy of subtraction. Since Quine rejects Kripke's requirements on using one's words to make assertions, he sees no reason to accept Kripke's conclusion that we can use our sentences to make assertions only if there is more to our linguistic behavior than jack-in-the-box reactions to stimuli.

One might believe that a proper analysis of our concept of meaning implies that Grice, Strawson, and Kripke's requirements for meaningfulness are correct, and hence that Quine cannot deny their requirements without replacing our concept of meaning with a different one, thereby simply ceasing to talk about meaning at all.

Quine's reply to this sort of objection is well known: we should not worry about preserving or capturing all features of our ordinary concept of meaning. We should focus instead on formulating clear, useful, and explanatory theories. What is less well known is that Quine explicitly endorses a method for this purpose. Following Carnap, he calls it the "method of explication." To *explicate* a linguistic expression *e* that one finds useful in some ways yet problematic in others is to decide to use, in place of *e*, a different linguistic expression *e'* that preserves and clarifies what one takes to be useful about *e* yet avoids what one takes to be the problems with *e* (Quine 1960, §53). In Quine's view, the proper question to ask is not whether his rejection of Grice, Strawson, and Kripke's requirements for using sentences to make assertions is faithful to our concepts of meaning, language use, and assertion. The proper question is whether Quine's rejection of those supposed requirements furthers our goals as inquirers.

If the answer to this latter question is "yes," then Quine's rejection of the requirements is vindicated, and the meaning-theoretic objections fall flat.

Many philosophers ignore or dismiss this reply, because they believe that Quine's method of explication presupposes his indeterminacy thesis and thereby begs the question against conceptual analysis. Following Carnap, however, Quine emphasizes that a synonymy constraint would be out of place for the sorts of paraphrases that matter to science, and to philosophy when it is viewed as continuous with science, "even if the notion of synonymy … were in the best of shape" (Quine 1960, p. 208; see also p. 159). If we seek above all to provide the best explanations of observed phenomena, we should use our best current explications of terms previously in use and not concern ourselves with conceptual analyses of them, even if we suppose that synonymy is an objective relation. This methodological point stands on its own. It needs no support from Quine's indeterminacy thesis. In fact, the direction of support is the other way around: Quine's case for his indeterminacy thesis depends on his explication of meaning as what is determined by the totality of speech dispositions.

One might think that there is a better explication of meaning than Quine's, an explication that does not have the result that translation is indeterminate. Perhaps there is. But the possibility of such an explication is no objection to Quine's argument that if we explicate meaning as he proposes, synonymy is not an objective relation, and meaning is not normative. Nor is it an adequate response to Quine's charge that many philosophers commit the fallacy of subtraction. For Quine is surely within his rights to point out that if we adopt his explication of meaning, we can coherently reject Grice, Strawson, and Kripke's requirements for meaningfulness while still acknowledging that we use our sentences to make assertions.

## 5.4 Language Use and Searle's Objection

To remind us what it is to use our words to make assertions, and to try to convince us that such language use is compatible with his indeterminacy thesis, Quine displays particular uses of sentences and invites us to acknowledge that they are uses of sentences to make assertions. Recall, for instance, that when Quine introduces his disquotational account of truth in §6 of *Word and Object*, he simply disquotes and uses his sample sentences, 'Brutus killed Caesar' and 'The atomic weight of Sodium is 23'. His method here is the same as in his earlier paper "Notes on the Theory of Reference," in which he emphasizes that Tarski's method of defining truth serves to endow ... 'true-in-$L$' ... with every bit as much clarity, in any particular application, as is enjoyed by the particular expressions of $L$ to which we apply [it]. Attribution of truth in particular to 'Snow is white', for example, is every bit as clear to us as attribution of whiteness to snow. (Quine 1953c, p. 138)

Here and in other passages, Quine elucidates his conception of language use by putting language use on display in a way that reminds us of its peculiar clarity and transparency.

In "Ontological Relativity," Quine tries once again to elucidate his resolutely minimalist notion of language use. This time he displays language use as a response to a sophisticated new version of the Grice–Strawson challenge to the coherence of his indeterminacy thesis. He notes that the indeterminacy thesis applies to one's own language, and that this seems to imply that "there is no difference on any terms, interlinguistic or intralinguistic, objective or subjective, between referring to rabbits and referring to rabbit parts or stages" (Quine 1969a, p. 47). In the objector's voice, he writes, "Surely this is absurd, for it would imply that there is no difference between the rabbit and each of its parts or stages" (Quine 1969a, pp. 47–48).

Quine states the supposed absurdity in this way precisely in order to show that there is some mistake in the reasoning that leads to it. To recognize the absurdity of the conclusion that there is no difference between the rabbit and each of its parts or stages is also, and thereby, to recognize that we can and do deny that *there is no difference between the rabbit and each of its parts or stages* – in just those words.

John Searle argues that we cannot deny this, in just those words, without committing ourselves to rejecting Quine's indeterminacy thesis.[1] According to Searle, we must presuppose that the references of our own

[1] Searle 1987, pp. 130–133. In a similar vein, though without claiming that Quine's position is incoherent, Barry Stroud notes that it is natural to think that "the sentence 'The word "rabbit" refers to rabbits' will explain the reference of that word to us only if we know what rabbits are. And someone's uttering that disquotational sentence would amount to saying that the word 'rabbit' refers to rabbits only if he is referring to rabbits in using (not mentioning) that word in that sentence" (Stroud 2000, p. 161). This natural thought poses a serious difficulty for Quine, since his indeterminacy thesis applies to one's own words, not just those of other speakers, and undermines what appears to be our presupposition that our word 'rabbit' determinately refers to rabbits. How then can we use our word 'rabbit' to say that it refers to rabbits? Many other philosophers endorse something like Searle's reasoning, which is now deeply entrenched in the literature. For instance, Ernie Lepore and Kirk Ludwig reject Donald Davidson's version of Quine's thesis that meaning and reference are indeterminate as incoherent on the grounds that an interpreter "cannot regard his own language as subject to the same indeterminacy that he is supposed to think of his subject's language as being subject to" (Lepore and Ludwig 2005, p. 385).

words are determinate when we use them to say such things as that rabbits are distinct from rabbit parts or rabbit stages. And Searle believes that Quine himself acknowledges the necessity of this presupposition when Quine notes the absurdity of the conclusion that there is no difference between the rabbit and each of its parts or stages. For at that critical point in his argument, Quine seems to acknowledge that we cannot simultaneously take ourselves to use our words to make assertions and reject the determinacy of reference. If Searle is right, Quine's resolutely minimalist account of language use is internally incoherent.

There is a striking structural similarity between Searle's objection and the other meaning-theoretic objections discussed above. In each case some requirement for using our words to make assertions is laid down; once the requirement is laid down, it is argued that Quine's account of meaning violates it and thereby implies that we cannot make assertions. This structural similarity should lead us to wonder whether Searle's argument also commits something like the fallacy of subtraction. But in Searle's case, what might the fallacy be?

It would not help to point out that in Quine's view, even though reference is indeterminate, every acceptable translation of my words yields some assignment or other of references to them. For if this remark is supposed to bear directly on our understanding of language use, it threatens to spoil our sense of the peculiar *transparency* of language use, including the use on display when we recognize the absurdity of saying that there is no difference between the rabbit and each of its parts or stages. To take this threat seriously is to endorse Searle's central assumption that we can use our words to make assertions only if reference is determinate. And if we endorse Searle's central assumption, we will see no way to resist Searle's conclusion that Quine's indeterminacy thesis is incoherent.

Despite its intuitive appeal, however, Searle's central assumption is mistaken. Quine explicates reference as satisfaction, defined in the first instance disquotationally. Thus explicated, reference is exclusively a relation between words and things that we define by using our words. Reference in this sense is not at issue when one uses one's nonsemantic vocabulary in the resolutely minimalist sense that is pragmatically indispensible to Quine's philosophy. Insofar as we are thus using, and not also mentioning, a given string of nonsemantic words, no particular relation between those words and things is asserted, presupposed, or implied.

There is textual evidence that this is exactly how Quine would respond to the sort of problem that Searle raises. Quine apparently thinks we can block the absurd implication that there is no difference between the rabbit and each of its parts or stages by displaying our language use without talking of reference at all. He attempts this in the following passage:

> Toward resolving this quandary, begin by picturing us at home in our language, with all its predicates and auxiliary devices. This vocabulary includes "rabbit," "rabbit part," "rabbit stage,"...; also the two-place predicates of identity and difference, and other logical particles. In these terms we can say in so many words that this is ... a rabbit and that a rabbit part, this and that the same rabbit, and this and that different parts. *In just those words.* (Quine 1969a, p. 48)

Quine's aim here is to remind us that we use our words to say, for instance, that this is a rabbit and that a rabbit part, this and that the same rabbit, and this and that different parts. By reminding us of indexical applications of our sentences, Quine also reminds us that our use of words is responsive to publicly observable conditions, and thereby hints at a final, crucial point: his minimalist account of language use in no way conflicts with his view that meaning is public and indeterminate.

## 5.5   Epistemology Again

Let us revisit the objection that what Quine tells us about epistemology from the standpoint of our own evolving theory conflates a speaker's taking a given assertion to be justified with its being justified. How might Quine respond to this objection? Let us grant, for the reasons just explained, that Quine can assume that we are each always in a position to express conditions under which our sentences are true by simply using them. This is not to equate their truth with our current beliefs about whether they are true. After I have made a particular judgment, I may later look back on that judgment and regard it as mistaken. From the standpoint of my current best theory, I can criticize another theory, including one of my own past best theories. At the time of adopting a theory, of course, I cannot distinguish from within my subjective point of view between having justification and taking myself to have justification. But when I reflect on one of my previous judgments, I may see something that I overlooked at the time, something that from my current point of view seems to me to undermine my previous confidence that my judgment was justified. Perhaps I come to think, for example, that while I took myself to have been exercising the best kind of scientific judgment on a given occasion, in fact I was not doing so. One need not believe that there are universal, context-independent standards for exercising scientific judgment in order to accept the possibility of this kind of retrospective assessment of whether a judgment is justified.

One might think that this reply does not yield an objective purchase on the notion of justification. For the best we can do, according to the Quinean position just stated, is to judge justification from our best current theoretical point of view. To evaluate a past judgment, we would first have to choose a translation *T* from our past idiolect to our present one and then evaluate past judgments, as mapped by *T* into our current idiolect, by our current standards. At any given time, we have no way to distinguish between our current inclination to take a judgment as justified, on the one hand, and its being justified, on the other.

To regard this objection as decisive one must suppose that there is some privileged standpoint from which to evaluate whether or not particular claims are justified – a standpoint on justification that is independent of any particular theory that we adopt. In Quine's view, however, there is no such standpoint. Scientific method, unsupported by ulterior controls, is "the last arbiter of truth" (Quine 1960, p. 23) and we have no grip on scientific method apart from our own evolving scientific theories. Hence, for Quine it is only from the standpoint of our best current theories that we can judge truth "as earnestly and absolutely as can be" (Quine 1960, p. 25).

This aspect of Quine's philosophy – the uncompromising core of his scientific naturalism – can seem puzzling. For if there is no legitimate standpoint on justification that is independent of any particular scientific theories we accept, then there are no substantive general principles for evaluating or justifying our acceptance of particular scientific statements or the theories of which they are a part. How then can an inquirer's acceptance of particular statements and theories be epistemically reasonable?[2]

Quine's response to this question combines an application of his method of explication with an application of his method of elucidation by subtraction.[3] To get on with our inquiries, he observes, we need a

[2] This question is similar to one raised by Gideon Rosen, who suggests that Quine has no good answer to what Rosen calls the "Authority Problem for Naturalized Epistemology" – the problem of giving "some sort of principle for telling the authoritative practices from the rest" (Rosen 1999, p. 471).
[3] Quine also develops a theory of how our sentences are associated with sensory stimulation by the mechanism of conditioned response. But such descriptions do not convey what it is to use one's sentences to accept theories or to justify, defend, or revise statements from the standpoint of a theory one has accepted. When Quine tries to convey his minimalist understanding of what it is to use one's sentences in these familiar ways, his task is not to theorize about it, but to elucidate it.

practical grasp of when it is appropriate to adopt a theory and what it is to justify, defend, and revise our assertions from the standpoint of a given theory. The best way to attain such a grasp is to immerse oneself, by study and practice, in the relevant scientific disciplines. Such immersion yields a practical and theoretical grasp of the disciplines that in no way depends on one's reflection about or acceptance of any traditional philosophical theory of justification. Quine concludes that an inquirer who has immersed herself in some scientific disciplines and wishes above all to get on with her inquiries has no need for a fully general, discipline-independent account of what it is for her acceptance of a given theory or statement in any of her disciplines to be justified.

One might challenge this conclusion on the grounds that an inquirer who starts by fully endorsing some scientific disciplines she has learned might subsequently be led by her own skeptical reflections to conclude that by the standards implicit in those scientific disciplines, and contrary to what she used to assume, she actually has no justification for any of her scientific assertions. Such an inquirer might arrive at the conclusion that to understand how she can justify any assertion she needs to have a fully general understanding of how human knowledge is possible at all – an understanding that she despairs of ever attaining.[4] That an inquirer could arrive at a skeptical conclusion in this way may appear to discredit Quine's conclusion that an inquirer has no need for a fully general, discipline-independent account of what it is for her acceptance of a given theory or statement in any of her disciplines to be justified.

This challenge rests on the assumption that there are standards – in particular, certain requirements for epistemological justification – that traditional philosophers see in our scientific practices but that Quine's minimalist account of those practices fails to capture. As I noted in §5.3 above, Quine counters other such challenges to his philosophy by judicious applications of his method of explication. In epistemology, or what little is left of it when its questions are posed from the standpoint of an inquirer who is engaged in scientific inquiry, Quine's scientific naturalism commits him to explicating and elucidating epistemic terms in a resolutely minimalist way that preserves applications of the epistemic terms that inquirers find clear and useful without leaving inquirers vulnerable to traditional skeptical arguments. Like inquiry itself, this task may never be

---

[4] According to Stroud, if we take the traditional skeptic's reasoning seriously we will come to realize that "we want an account of our knowledge of the world that would make all of it intelligible to us all at once," but that such an account is impossible, since "we cannot consider all our knowledge of the world all at once and still see it as knowledge" (Stroud 1984a, p. 551).

completed. New methods in science may create new vulnerabilities to old skeptical arguments that call for new explications. Once we replace the method of conceptual analysis with the method of explication, however, there is no general reason to worry that our inquiries can be decisively undermined from within by traditional skeptical doubts.[5] Hence, if our primary goal is to get on with our inquiries, as Quine assumes, we are free to explicate the phrase "epistemically reasonable" in such a way that our judgments and inferences can be epistemically reasonable even if there are no substantive general principles for evaluating them.[6]

One might object that this explication abandons our concept of what it is to be epistemically reasonable and replaces it with a radically different one that has nothing to do with epistemology. But this objection can appear to us to discredit Quine's epistemology only if we cling dogmatically to the claim that if we abandon the traditional epistemologist's assumption that there are substantive general principles for evaluating and justifying assertions, we lose our grip on what it is to justify, defend, and revise our assertions from the standpoint of a given scientific discipline. To challenge this claim, all Quine can do is invite us to subtract the assumption that there are substantive general principles for evaluating and justifying assertions from the practical and theoretical grasp of what it is to justify, defend, and revise assertions in a given scientific discipline that we achieve by study and practice of that discipline. Subtracting this assumption from our practical and theoretical grasp of what it is to justify theories or statements in the scientific disciplines we have learned or developed, we get Quine's resolutely minimalist understanding of justification as the difference.

The resulting account of justification will not satisfy an epistemologist who wishes above all to capture and preserve her traditional epistemological preconceptions, including her assumption that there are substantive

---

[5] Dirk Koppelberg offers a roughly similar response on Quine's behalf to traditional skepticism, though without mentioning Quine's methods of explication and elucidation by subtraction. Koppelberg writes, "What a [Quinean] naturalist denies is the skeptic's presupposition of the intelligibility of a completely external perspective on our knowledge.... If we resist the temptation to adopt an external standpoint, in favor of trying to explain our knowledge from within our evolving and changing theory of the world, the significance of philosophical skepticism might yet decline. When it amounts to the insight that our traditional notions of knowledge and objectivity stand in need of revision, naturalism can be seen to offer a promising way with this task at hand" (Koppelberg 1990, p. 210). In Quine 1990b, Quine expresses his "thanks and admiration" for Koppelberg's analysis of his epistemological position.

[6] Consider in this light the last sentence of "Two Dogmas of Empiricism": "Each man is given a scientific heritage plus a continuing barrage of sensory stimulation; and the considerations which guide him in warping his scientific heritage to fit his continuing sensory promptings are, where rational, pragmatic" (Quine 1953b, p. 46).

general principles for justifying assertions. But the objections of such an epistemologist are no more challenging to Quine's efforts to elucidate our practical and theoretical grasp of what it is to justify theories or statements in the scientific disciplines we have learned or developed than are Grice, Strawson, Kripke, or Searle's objections to Quine's efforts to elucidate our use of sentences to make assertions. The two sorts of objections are, indeed, at root identical.[7] Quine's response to them is to reject their assumptions and to rely on our pragmatically prior grasp of what it is to assert and justify theories and sentences from the standpoint of a scientific discipline we endorse.

## 5.6   The Last Word

For just this reason, Thomas Nagel sees Quine's philosophy as a form of subjectivism that "shrink[s] from the apparently audacious pretensions of human thought and tend[s] to collapse its content into its grounds" (Nagel 1997, p. 7). Nagel observes that we treat the results of our best current first-order theorizing as the last word – as absolutely objective and unconditionally authoritative for us right now – unless and until we see reason to revise the results, and thereby to arrive at a new result that we treat as the latest last word. Quine can easily paraphrase this preliminary observation into the resolutely minimalist language of his scientific naturalism. But Nagel goes further. He sees Quine's scientific naturalism (wrongly, I think, for reasons I have given above) as an attempt to underwrite the authority of our judgments with a psychological theory of our needs and linguistic dispositions. Against this, Nagel stresses (rightly, I think), "The authority of the most fundamental kinds of thought reveals itself only from inside each of them and cannot be underwritten by a theory of the thinker" (Nagel 1997, p. 26). Surprisingly, he goes on to theorize about what it is for our thoughts to be authoritative "from the inside." He claims, for instance, that we do not understand the authority of some of our thoughts unless we can give substantive content to the truism that "to reason is to think systematically in ways anyone looking over my shoulder ought to be able to recognize as correct" (Nagel 1997, p. 5). And he argues that this truism has substantive content only if certain rationalist

---

[7] Recall Quine's claim that the dogma that there is an analytic–synthetic distinction and the dogma of reductionism "are, indeed, at root identical" (Quine 1953b, p. 41). The two dogmas that Quine identifies are of a piece with the meaning-theoretic and epistemological objections to Quine's philosophy that I have considered in this chapter.

assumptions are true. More generally, in his view, we can reason systematically, debate our reasons with others, and use our language to make assertions that count for us now as the last word on a given topic only if our sentences have objective contents or meanings and our judgments are justified by universally binding reasons. Nagel concludes that since Quine tries to do without these rationalist assumptions, he fails to explain the phenomenon of the last word.

My reply to Nagel's criticisms of Quine is similar to my replies to the criticisms that I discussed above. The apparent appeal of Nagel's rationalistic requirements on using one's sentences to express one's last word on a given topic is an illusion sustained by a flawed philosophical methodology, one that places conceptual analysis at center stage, and hence fails to keep truth in focus. The best way to avoid such illusions is to replace the method of conceptual analysis with the method of explication, and to keep one's philosophy focused on truth, not on concepts or justification. To reason and to judge, "we can never do better than occupy the standpoint of some theory or other, the best we can muster at the time" (Quine 1960, p. 22). Hence, it is fruitless to look for universally binding justifications for our reasoning and judging, beyond those reasons we already offer when we apply scientific method to arrive at our best current theories. We must therefore reject Nagel's rationalistic requirements on reasoning and judging, along with all philosophical dreams of a perspective on reasoning and judging that is somehow higher or firmer than our own current theories. Only when our rejections of such misguided philosophical theorizing are complete can we attain a clear view of what it is to take a bit of discourse as the last word – to judge truth "as earnestly and absolutely as can be" (Quine 1960, p. 25). In short, it is Quine, not Nagel, who gets the last word.

# CHAPTER 6

# Reading Quine's Claim That Definitional Abbreviations Create Synonymies

In §2 of "Two Dogmas of Empiricism" Quine claims that when we explicitly introduce a new, previously meaningless expression, $A_s$, to serve as an abbreviation for a longer, already meaningful expression, S, we thereby create a "transparent" synonymy between $A_s$ and S (Quine 1953b, p. 26). In §3 and §5 of "Two Dogmas," Quine explains why he is skeptical of all *other* species of synonymy.

According to H. P. Grice, P. F. Strawson, Paul Boghossian, Scott Soames, and many others, Quine's claim that definitional abbreviations create synonymies is in tension with his skepticism about the notion of synonymy for other cases. I shall argue that this kind of criticism is based on a fundamental misunderstanding of Quine's reasoning in "Two Dogmas."

## 6.1 A First Look at the Context for Quine's Claim

In §1 of "Two Dogmas of Empiricism" Quine grants that first-order logical truths are analytic and asks how to extend this partial characterization of the set of analytic sentences to include such supposedly analytic sentences as "Bachelors are unmarried." He accepts, provisionally, that a sentence is in the wider class if "it can be turned into a logical truth by putting synonyms for synonyms" (Quine 1953b, p. 23), but notes that this characterization of analyticity "lean[s] on a notion of 'synonymy' which is no less in need of clarification than analyticity itself" (Quine 1953b, p. 23).

In §§2–3 of "Two Dogmas" Quine considers various ways of trying to clarify the relevant sense of the term 'synonymy' and its cognates, as these terms appear in his assumed, preliminary characterization of the wider class of analytic sentences.

In §2, Quine considers the initially tempting idea that the relevant synonymies rest on, and are therefore explained by, definitions. Against this idea, he argues that some of the synonymy relations that are supposed to be relevant to the wider class of analytic statements hold between expressions, such as 'Bachelor' and 'unmarried man', that were already in use prior to any explicit definitions we may have subsequently given them. In such cases, as well as cases in which we seek to "improve on" a term already in use "by refining or supplementing its meaning," our definitions do not clarify the synonymy relation relevant to analyticity, since they are supposed to be faithful to prior and independent relations of sameness or similarity of meaning between a term we define, the *definiendum* (such as 'Bachelor'), and the terms in which we define it, the *definiens* (such as 'unmarried man').

Immediately following these observations, Quine makes his claim about definitional abbreviation:

> There does, however, remain still an extreme sort of definition which does not hark back to prior synonymies at all; namely, the explicitly conventional introduction of novel notations for purposes of sheer abbreviation ... the definiendum becomes synonymous with the definiens simply because it has been created expressly for the purposes of being synonymous with the definiens. Here we have a really transparent case of synonymy created by definition; would that all species of synonymy were as intelligible. For the rest, definition rests on synonymy rather than explaining it. (Quine 1953b, p. 26)

Quine concludes that the initially tempting thought that synonymies rest on, and are therefore explained by, definitions is incorrect: "Definition does not hold the key to synonymy" (Quine 1953b, p. 27). Definition therefore cannot help us extend the first, partial characterization of the set of analytic sentences as the set of first-order logical truths, to obtain a clear characterization of a wider class of analytic truths that includes such supposedly analytic sentences as "Bachelors are unmarried."

## 6.2   The Grice–Strawson Criticism

H. P. Grice and P. F. Strawson object to Quine's claim in the above passage about definitional abbreviation, as follows:

> If we are to take these words of Quine seriously, then his position *as a whole* is incoherent. It is like the position of a man to whom we are trying to explain, say, the idea of one thing fitting into another thing, or two things fitting together, and who says: "I can understand what it means to say that one thing fits into another, or the two things fit together, in the case where one was specifically made to fit the other; but I cannot understand what it means to say this in any other case." Perhaps we should not take Quine's words too seriously. But if not, then we have the right to ask him exactly

what state of affairs he thinks is brought about by explicit definition, what relation between expressions is established by this procedure, and why he thinks it unintelligible to suggest that the same (or a closely analogous) state of affairs, or relation, should exist in the absence of this procedure. For our part, we should be inclined to take Quine's words (or some of them) seriously, and reverse his conclusions; and maintain that the notion of synonymy by explicit convention would be unintelligible if the notion of synonymy by usage were not presupposed. (Grice and Strawson 1956, pp. 152–153)

It is striking that Grice and Strawson equate taking Quine's words seriously with reading his claim that definitional abbreviations create transparent synonymy in a way that is incompatible with his skepticism about the notion of synonymy by usage. They mention the possibility that Quine might be able to explain "what state of affairs he thinks is brought about by explicit definition," but they don't pause to investigate this. I would have thought, however, that to take Quine's words seriously is to try to understand them in the way that Quine understood them. Instead of making this effort, however, they "reverse his conclusions" by pressing the criticism that "the notion of synonymy by explicit convention would be unintelligible if the notion of synonymy by usage were not presupposed."

Many others have raised similar criticisms. Paul Boghossian, for instance, writes,

> [Quine's] skepticism about synonymy has to boil down to the following somewhat peculiar claim: Although there is such a thing as the property of synonymy; and although it can be instantiated by pairs of tokens of the same orthographic type; and although it can be instantiated by pairs of tokens of distinct orthographic types, provided that they are related to each other by way of an explicit stipulations; it is, nevertheless, in principle impossible to generate instances of this property in some other way, via some other mechanism. (Boghossian 1996, p. 372)

And in a similar vein, Scott Soames writes that

> Quine ... seems to have forgotten that the issue that is central to his overall argument is not how synonymies get created, but whether the notion of synonymy – i.e. sameness of meaning – makes sense. His position is that it doesn't. But if it doesn't, then to grant that explicitly stipulated synonyms are genuinely synonymous is to say something inconsistent with his overall conclusion. It is telling, I think, that even Quine's dedication to his larger, negative, argumentative purposes was not enough to prevent a glimpse of the denied truth from breaking through. (Soames 2003, pp. 364–365, fn 8)

What Soames does not mention is that the burden of §2 of "Two Dogmas" is not to establish that synonymy is a doubtful notion, but to show that definition does not clarify the notion of synonymy, except in the special case of explicitly conventional abbreviations. How then could Soames have arrived at the conclusion that Quine "seems to have forgotten that the issue that is central to his overall argument is … whether the notion of synonymy – i.e. sameness of meaning – makes sense"? The answer, I think, is that like Grice, Strawson, and Boghossian, Soames believes to accept that there are *any* clear cases of synonymy is thereby also to accept that a fully general notion of synonymy makes sense, and is therefore incompatible with doubting that the notion of synonymy makes sense in general, for any two expressions of a language.

Grice, Strawson, Boghossian, and Soames appear to think that there's something about the very idea of synonymy that Quine is committed to rejecting. They therefore see Quine's insistence that he finds *some* cases of synonymy clear as incompatible with his skepticism about the notion of synonymy for other cases. What exactly do they take the incompatibility to be?

To answer this question it helps to consider a related criticism articulated by William Lycan, who says that Quine's "concession" that definitional abbreviations create synonymies

> is larger and more damaging than perhaps Quine realized in making it. For it seems to allow that there are analyticity-generating synonymies and so genuine instances of full-fledged analyticity, even if they are few. … The idea of adventitious stipulated definition *as it would have to be construed for the purpose of saving analyticity* must presuppose the notion of strict, analyticity-generating synonymy. (Lycan 1991, p. 122)

The problem, according to Lycan, is that in order to save analyticity, stipulated definitions would have to be construed in such a way that "[they] and their consequences *cannot be* false, and thus are necessarily true" (Lycan 1991, p. 125). Thus understood, however, Quine's claim that definitional abbreviations create synonymies is incompatible with Quine's well-known rejection of the view that there are some statements that *cannot* be false, in the sense that they are *necessarily* true, where *being necessarily true* is understood as an objective modal property of the statement.

Let us call this *the modal criticism*. It may be at least part of what Grice, Strawson, Boghossian, and Soames had in mind, and would explain why Soames, for instance, says that "to grant that explicitly stipulated synonyms are genuinely synonymous is to say something inconsistent with [Quine's] overall conclusion." In response to the modal criticism, previous commentators have stressed that according to Quine, definitional abbreviations are both transient and not guaranteed to be true. (See Harman 1967, p. 132, 140; Ebbs 1997, pp. 135–136; Russell, G. 2014, p. 190.) While this observation is correct and deflects the modal criticism, it is not based on close reading of Quine's reasoning in §§1–2 of "Two Dogmas." I show in §6.3 that the modal criticism is irrelevant to Quine's reasoning in §§1–2 of "Two Dogmas," which does not rely on the additional, though correct, point that definitional abbreviations are transient: when Quine says that definitional abbreviations create synonymies, he is granting only that there's a kind of synonymy that can hold between two expressions by dint of how we use them. He is not granting that such synonymies and their consequences cannot be false, and thus are necessarily true.

There is, however, another, more challenging criticism that is suggested by what Grice, Strawson, Boghossian, and Soames say. The criticism is that if explicitly conventional definitional abbreviation create synonymies by usage, then, contrary to what Quine says, the notion of synonymy by usage must also make sense for pairs of expressions neither of which is an explicitly conventional definitional abbreviation of the other. I will call this *the synonymy-by-usage criticism*. In §§6.4–6.6 I explain why the synonymy-by-usage criticism is also mistaken.

## 6.3   Quine's Commitments and Methods in "Two Dogmas"

To see why the *modal* and *synonymy-by-usage* criticisms are mistaken, one needs to be aware of the philosophical commitments and methods of inquiry that guide Quine's reasoning in "Two Dogmas."

### Scientific Philosophy

Following his teacher and friend Rudolf Carnap, Quine is committed to a revolutionary philosophical orientation that he calls "empiricism," or "scientific philosophy" (Quine 1949, p. 156), according to which it is only from within science that reality is to be identified and described (Carnap 1934, Quine 1981, p. 21). Quine employs Carnap's own preferred method for revealing the emptiness of a philosopher's words – a method partly inspired by Wittgenstein's pronouncement, in his *Tractatus Logico-Philosophicus*, that the correct method in philosophy [is] to say nothing except what can be said, i.e. propositions of natural science – i.e. something that has nothing to do with philosophy – and then, whenever someone else wanted to say something metaphysical, to demonstrate to him that he had failed to give a meaning to certain signs in his propositions. (Wittgenstein 1921: 73–74)

Quine's aim in "Two Dogmas" is to demonstrate that "empiricists," or "scientific philosophers," including, of course, Carnap, have failed to explain their term 'analytic' in a language suited for and used in the mature natural sciences. As Quine later explained, his doubts about Carnap's analytic–synthetic distinction were an expression of "the same sort of attitude, the sort of discipline that Carnap shared and that I owed, certainly, in part to Carnap's influence: I was just being more carnapian than Carnap in being critical in this question" (Quine 1994, p. 228). In §4 of "Two Dogmas," after canvassing "explanations of analyticity known to Carnap and his readers," Quine concludes that

> for all its a priori reasonableness, a boundary between analytic and synthetic statements simply has not been drawn. That there is such a distinction to be drawn at all is an unempirical dogma of empiricists, a metaphysical article of faith. (Quine 1953b: 37)

Quine's strategy for making the case that no boundary between analytic and synthetic has been drawn in a language suited for and used in the mature natural sciences is to demonstrate to Carnap and others that none of their attempts to explain the terms "analytic" and "synthetic," as they wish to apply them, is successful. From his demonstrations of the failures of these attempts, Quine concludes – by an application of Carnap's own Wittgenstein-inspired methodology – that the unexplained terms should be dropped from scientific philosophy.

### *Extensional Constraints on a Successful Explication of 'Analytic'*

To apply this strategy, Quine begins with sentences that are analytic "by general philosophical acclaim" (Quine 1953b, p. 22), and tries to find a clear, scientific explication of 'analytic' that applies to these sentences. For Quine's purposes, the constraints on a successful explication of 'analytic' are that

(1)   Sentences that are analytic "by general philosophical acclaim" should be analytic according to the explication; and

(2)   A sentence should be analytic according to the explication only if it is true (Quine 1953b, p. 34).

An explication that satisfies constraints (1) and (2) may be purely extensional, and hence say nothing about whether an analytic sentence is one that cannot be revised without a change in meaning. One might therefore think that the constraints are too weak. But any acceptable explication of analytic must *at least* satisfy constraints (1) and (2), and this implies that constraints (1) and (2) are sufficient for Quine's critical purposes. For if, as Quine argues, there is no scientifically respectable *extensional* explication of 'analytic' that satisfies constraints (1) and (2), then a fortiori there is no scientifically respectable *intensional* explication of 'analytic' that both satisfies constraints (1) and (2) and implies that an analytic sentence cannot be revised without a change in meaning.

This is the heart of my response to the modal criticism I described in §6.2: Quine's minimal requirement in "Two Dogmas" is not that scientific philosophers who wish to use the term 'analytic' be able to explain or clarify their assumption that an analytic statement is necessarily true, but simply that they provide a scientifically respectable *extensional* explication of 'analytic' that satisfies constraints (1) and (2). That this is all that Quine requires becomes clear when one examines the first two steps of his effort to clarify the term 'analytic'.

## Quine's First Step: Logical Truths Are Analytic

As I briefly noted above, Quine's first main step in §1 of "Two Dogmas" is to grant (for the sake of his inquiry) that first-order logical truths are among the acclaimed 'analytic' sentences. He defines the first-order logical truths in terms of truth and substitution, as follows:

> If we suppose the prior inventory of *logical* particles, comprising 'no', 'un-', 'not', 'if', 'then', 'and', etc., then in general a logical truth is a statement which is true and remains true under all reinterpretations of its components other than the logical particles. (Quine 1953b, pp. 22–23)

This seemingly casual account of logical truth is a summary of an account of logical truth that Quine develops in full technical detail in his books *Mathematical Logic* (Quine 1940), *Elementary Logic* (Quine 1941), and *Methods of Logic* (Quine 1950, 1959, 1972, 1982). Properly and fully articulated, the relevant "reinterpretations" are characterized syntactically (Quine 1982, chapters 26 and 28), and the notion of truth for a given language is defined by Tarski's methods (1953c). For rich enough languages, the resulting "substitutional" account of first-order logical truth is known to be equivalent to the standard model-theoretic account of first-order logical truth (Kleene 1952, theorem 35; Quine 1986f; Ebbs 2015).

In taking this first step, Quine follows Carnap, who proposed it himself in discussions with Tarski and Quine in 1941 (Frost-Arnold 2013, p. 156). Quine therefore had good reason to take his extensional account of first-order logical truth as uncontroversial among the scientific philosophers of his day who wished to use the term 'analytic'. But Quine's acceptance of Carnap's proposal does not signal his acceptance or understanding of the idea that logically true sentences are "true in virtue of meaning," or non-empirical. As he explains in *Word and Object*:

> Those who talk confidently of analyticity have been known to disagree on analyticity of truths of arithmetic, but are about unanimous on that of the truths of logic. *We who are less clear on the notion of analyticity may therefore seize upon the generally conceded analyticity of the truths of logic as a partial extensional clarification of analyticity; but to do so is not to embrace the analyticity of the truths of logic as an antecedently intelligible doctrine.* (Quine 1960, p. 65, note 3; my emphasis)

Quine's point is just that the class of supposed analytic sentences of a given language *L* contains the set of first-order logical truths of *L*, where these truths are characterized extensionally, in terms of truth and substitution.

When Quine says that logical truths are analytic, he therefore is not granting that they *cannot be* false, and thus are necessarily true, where *being necessarily true* is understood as an objective modal property of a logical truth. His explication therefore fails to capture any modal force to the claim the statement is analytic. Contrary to the modal criticism described in the previous section, the same goes for Quine's explication of synonymy created by definitional abbreviation, as I shall now try to show.

### Quine's Second Step: 'Analytic' = 'Can Be Turned Into a Logical Truth by Putting Synonyms for Synonyms'

As I also briefly noted above, Quine's second step in §1 of "Two Dogmas" is to propose that a sentence is in the wider class of analytic truths if "it can be turned into a logical truth by putting synonyms for synonyms" (Quine 1953b, p. 23). Unlike 'analytic', of course, the ordinary English words 'synonym' and 'synonymy' have many useful everyday applications. If we seek to clarify the technical term 'analytic', as it is used by scientific philosophers, such as Carnap, however, we need a precise technical characterization of a relevant notion of 'synonym' and 'synonymy'. The problem is that as these terms are ordinarily used, what counts as a synonym of a given word in a given context depends on what sorts of "meaning" and "sameness of meaning" matter for our purposes in the context. Are we concerned about emotional, associative, or cognitive meaning? If we are concerned with cognitive meaning, do we characterize cognitive meaning in terms of verification conditions, informational content, or truth conditions, and, if any of these, which particular senses of them? To specify the wider class of analytic truths, we would need a technical notion of synonymy that is not multiply ambiguous in this way, one that is unambiguous and can be clarified using only terms taken from the sciences, including, of course, the science of logic.

On this issue, also, Quine follows Carnap's lead. Carnap characterizes the idea of an analytic statement as one whose truth value is "not dependent upon the result of observations and therefore can be given before any relevant observations are made" (Carnap 1942, p. 61). He emphasizes however that this and similar explanations "merely circumscribe what it is we are looking for" (Carnap 1942, p. 63), and that "our problem will be to transform this vague characterization into a precise definition" (Carnap 1942, p. 60). In particular, Carnap agrees that to specify the wider class of analytic truths, we would need a technical notion of synonymy that is unambiguous and explained using only terms taken from the sciences.

Quine's question in §2 is whether we can use definition, one of the established tools of logic, to clarify the supposed technical notion of synonymy that would be needed to explicate the wider class of analytic sentences of a language. His answer is that with the sole exception of definitional abbreviations, the synonymies we would need to clarify in order to specify the wider class of analytic truths are not "created" by definitions, but presuppose independent and prior relations of synonymy that do not hinge on our adoption of definitions. Just as in the case of logical truth, however, when Quine says that definitional abbreviations create synonymies, he is not granting that "[they] and their consequences *cannot be* false, and thus are necessarily true," as Lycan's modal criticism assumes. As I shall explain in §6.4, Quine grants only that definitional abbreviations institute new uses of expressions that imply that some statements are mere transcriptions of logical truths. The abbreviations have no greater modal significance for Quine than do logical truths, which Quine characterizes in purely extensional terms. The modal criticism is therefore mistaken.

## 6.4   Quine's Claim Explained

Let's turn now to the synonymy-by-usage criticism: if explicitly conventional definitional abbreviation creates synonymies by usage, then, contrary to what Quine says, the notion of synonymy by usage must also make sense for pairs of expressions neither of which is an explicitly conventional definitional abbreviation for the other. To see why this criticism is also mistaken, one must understand the sense in which Quine grants that a definitional abbreviation "creates" a synonymy of the sort needed to define the wider class of analytic truths. This is easy to explain in scientific terms, given Quine's assumption (from §1 of "Two Dogmas") that

(a)   First-order logical truths are analytic,

and Carnap's proposal that we define "Sentences $S_1$ and $S_2$ are synonymous" in term of analyticity as follows:

(b)   Sentences $S_1$ and $S_2$ are synonymous if and only if '$S_1 \leftrightarrow S_2$' is analytic (L-true). (Carnap 1937, pp. 176–177, p. 290; Carnap 1939, p. 15; Carnap 1947, pp. 14–15)

Although Quine does not explicitly affirm (b) in §2 of "Two Dogmas," his intended readers – fellow empiricists, or scientific philosophers – would surely have known of it. They would also have known that with some easy adjustments, one can apply (b) to define synonymy for open sentences, not only for closed ones. If $S_1$ and $S_2$ are two open sentences that contain all and only the same free variables, then analyticity of '$S_1 \leftrightarrow S_2$' amounts to the logical truth of the universal closure of '$S_1 \leftrightarrow S_2$'. Thus extended, we may apply (b) to define synonymy for two predicates $P_1(v_1,\ldots, v_n)$ and $P_2(v_1,\ldots, v_n)$ if we take an open sentence S containing $n$ free variables $v_1 \ldots v_n$ and simple general terms $F_1 \ldots F_m$ to represent a corresponding $n$-place predicate $\lambda x_1 \ldots x_n\, P(x_1,\ldots, x_n)$ that is constructed from $F_1 \ldots F_m$, truth functional connectives, and/or quantifiers.[1]

To simplify my discussion below, I will speak of synonymy of sentences, without distinguishing whether they are open or closed. I will assume, also, that synonymy of open sentences can be used to define synonymy of predicates that correspond to them in the way just explained.

By relying on (a) and (b) we may infer that *every sentence of our language is synonymous with itself*, as follows. Let $S$ be any sentence of our language. Since '$S \leftrightarrow S$' is a first-order logical truth, we may infer from (a) that

---

[1] Quine makes essentially the same point in §3 of "Two Dogmas" (Quine 1953b, p. 31).

'$S \leftrightarrow S$' is analytic. From this and (b), we may infer that $S$ is synonymous with itself.

As I read §2 of "Two Dogmas," Quine's question there is whether we can extend the reasoning just sketched so that it applies, also, to sentences of the form '$S' \leftrightarrow S$'. We can do this, he thinks, if $S'$ is a mere definitional abbreviation of S, so that '$S' \leftrightarrow S$' is a mere abbreviation of the logical truth '$S \leftrightarrow S$'. As Quine explains in *Mathematical Logic*, "In all discourse about theorems, defined notations are imagined expanded into primitive notations" (Quine 1940, p. 133). Thus suppose that $A_s$ is a new notation explicitly introduced as a conventional abbreviation for $S$ (where both $A_s$ and $S$ may be open sentences containing all and only the same variables). Then Quine's account of definitional abbreviation implies that '$A_s \leftrightarrow S$' is a logical truth (theorem) in the sense that it is just shorthand for (i.e., we imagine it expanded into) '$S \leftrightarrow S$', which is a logical truth (theorem) by the substitutional criterion that Quine sketches in "Two Dogmas," §1.

Here it is crucial to keep in mind that the kinds of definitional abbreviations in question – "the explicitly conventional introduction of novel notations for purposes of sheer abbreviation" – are those that take a previously meaningless term, $A_s$, and stipulate that it is to be mere shorthand for a longer, already meaningful expression, $S$. When we use $A_s$, we imagine it expanded into $S$. We introduce the shorter term only for convenience; given enough time and patience, we could do without it.

Thus conceived, definitional abbreviation is a central tool of logic, and is recognized as such by logicians (Russell 1903, p. 429; Frege 1914, pp. 207; Carnap 1942, p. 17; Quine 1936, pp. 78–79, Quine 1940, p. 47; Tarski 1941, pp. 33–36; Schoenfield 1967; Carnap 1958, p. 85). There is no question that particular acts of definitional abbreviation occur in logic all the time, and that this method of definition is therefore available to us when we try to clarify the boundary between analytic and synthetic sentences in scientific terms. The notion of synonymy that we can use it to clarify, with the help of (b), is very thin, however, and does not really go beyond Quine's first step of regarding first-order logical truths, explained in extensional terms, as analytic. In granting that definitional abbreviations create synonymies, Quine is granting no more than that such abbreviations license us to regard some sentences of the form '$A_s \leftrightarrow S$' as abbreviations of logical truths of the form '$S \leftrightarrow S$'.

In short, the synonymies Quine thereby finds "transparent" – synonymies between a novel notation $A_s$ that is introduced as definitional abbreviation for an already meaningful expression $S$ – are entirely explained by (b) together with the observation that '$A_s \leftrightarrow S$' is an abbreviation, or mere transcription, of '$S \leftrightarrow S$', which is a logical truth, and hence, by (a), analytic.

## 6.5   Textual Support for This Reading

When Quine discusses definitional abbreviation in §2 of "Two Dogmas," he does not say that synonymy can be defined in terms of analyticity. One might worry that this leaves my reading without much direct textual support. It is important to keep in mind, however, that unlike the Grice–Strawson reading of Quine's comments about definitional abbreviation in §2 of "Two Dogmas," the reading I offer does not attribute to Quine a claim that presupposes commitments he explicitly disavows in later sections of "Two Dogmas." I take this to be a point in favor of my reading. There are several additional reasons to prefer it, as well.

To begin with, whether in mathematics, logic, or philosophy, professional journal articles rest on a vast background of presumed shared knowledge and disciplinary skills. Quine clearly takes for granted, for example, that his readers have a good grasp of the resources of first-order logic, including Quine's substitutional characterization of logical truth, as well as familiarity with many related technical terms such as 'extensional' and 'necessary', where the latter is taken to rest on an explication of 'analytic'. Definition (b) is a close cousin of such explications of 'analytic' and Quine clearly assumes familiarity with it: at a crucial point in §3, Quine formulates a version of (b), namely, "Statements may be said simply to be cognitively synonymous when their biconditional (the result of joining them by 'if and only if') is analytic" (Quine 1953b, p. 32). In a footnote, seemingly intended for readers familiar with Carnap's views, he adds a revealing fine point, namely, that "The 'if and only if' itself is intended in the truth functional sense," and cites Carnap 1947, p. 14. This fine point of course does not by itself show that we should adopt (b); it simply reminds readers who already knew of (b) that according to Carnap, the biconditional in (b) is to be understood in extensional (i.e., truth functional) terms, and that (b) is therefore compatible with Quine's (ultimately unsuccessful) project in the first several sections of "Two Dogmas" of trying to provide extensional explications of analyticity and synonymy.[2]

---

[2] One might think that this textual evidence comes too late – in §3, not in §2, where, according to my reading, Quine already relies on it. When Quine introduces (b) in §3, he writes:

> The effort to explain cognitive synonymy first, for the sake of deriving analyticity from it afterward as in §1, is perhaps the wrong approach. Instead we might try explaining

One might wonder why Quine does not mention (b) in §2, where I claim he relies on it. This textual problem is partly answered by "Animadversions on the Notion of Meaning" (Quine 1949, henceforth "Animadversions"), notes for a paper Quine delivered at the University of Pennsylvania in December 1949, just one year before he presented "Two Dogmas" at an APA convention in Toronto. In "Animadversions," Quine focuses on three main topics, synonymy, analyticity, and significance, which he glosses as "possession of meaning" (Quine 1949, p. 152). He lays down the following explications of these terms:

> Significance = synonymy with self
> Analyticity = synonymy with 'o = o'
> Synonymy = analytic biconditional (Quine 1949, p. 153)

Quine notes that given these definitions, one can explain significance and synonymy in terms of analyticity. The last of these explications, "Synonymy = analytic biconditional," is a breezy expression of Carnap's definition of synonymy, summarized in (b) above.[3]

Following the above definitions, Quine tackles the idea of analyticity as "truth by virtue of meaning." He considers the proposal that "truth by virtue of meaning" amounts to "true by definition, i.e., definitional abbreviation of logical truth" (Quine 1949, p. 154), and distinguishes between two sorts of appeals to definition. According to the first, a definition is a "convention of abbreviation for convenience: imagine expanding always" (Quine 1949, p. 154; recall Quine's point in Quine 1940, that "in all discourse about theorems, defined notations are imagined expanded into primitive notations" (p. 133)). Such conventions yield "*Mere* logical truth" (Quine 1949, p. 154). According to the second, a definition is an "assertion of synonymy between two given forms." Here Quine clearly has

---

> analyticity somehow without appeal to cognitive synonymy. Afterward we could doubt-less derived cognitive synonymy from analyticity satisfactorily enough if desired. (Quine 1953b, p. 31)

One might infer from this passage that Quine did not view his remark about definitional abbreviation in §2 as resting on an account such as (b) of synonymy in terms of analyticity, as I suggest. In fact, however, what Quine states in the passage just quoted is that if one has an explanation of analytic that covers the wider class of analytic truths then one can use that explanation to clarify the notion of synonymy. This does not speak to the question of whether one can use (b) to explain the synonymy of a sentence with itself without having already provided an explanation of the wider class of analytic truths.

[3] Given (b), Quine's explication of "significance" is equivalent to Carnap's definition in Carnap 1950 of '*p* is a proposition' as '*p* or not *p*' – a definition Quine would have known about in December 1949, since Carnap sent Quine the manuscript version of Carnap 1950 in August 1949. See Carnap 1950, p. 210, and Letter 138: Carnap to Quine, August 15, 1949, in Creath 1990, pp. 415–417.

in mind not a "convention of abbreviation for convenience," but, as he puts it in "Two Dogmas," a definition that "hinges on a prior relation of synonymy."

There is additional support for my reading in Quine's papers "Truth by Convention" (Quine 1936) and "Carnap and Logical Truth" (Quine 1963a; written in 1954). In these papers Quine argues that any truth that appears to be established by a definitional abbreviation rests on a prior truth that is established independently of that definitional abbreviation. In "Truth by Convention," for instance, he writes:

> A definition, strictly, is a convention of notational abbreviation (cf. Russell 1903: 429).... From a formal standpoint the signs thus introduced are wholly arbitrary; all that is required of a definition is that it be theoretically immaterial, i.e., that the shorthand which it introduces admit in every case of unambiguous elimination in favor of the antecedent longhand.... Considered in isolation from all doctrine, including logic, a definition is incapable of grounding the most trivial statement; even 'tan $\pi$ = sin $\pi$/ cos $\pi$' is a definitional transformation of an antecedent self-identity, rather than a spontaneous consequence of the definition. (Quine 1936, pp. 78–79)

In "Carnap and Logical Truth," Quine repeats the point, as follows:

> Definition, in a properly narrow sense of the word, is convention in a properly narrow sense of the word. But the phrase 'true by definition' must be taken cautiously; in its strictest usage it refers to a transcription, by the definition, of a truth of elementary logic.... Even an outright equation or biconditional connecting the definiens and the definiendum is a definitional transcription of a prior logical truth of the form '$x = x$' or '$p \leftrightarrow p$'. (Quine 1963, p. 394)

In these passages Quine argues, in effect, that a definitional synonymy of the form '$A_s \leftrightarrow S$' is really just a definitional transcription of a logical truth, namely, '$S \leftrightarrow S$', which, as we saw, given (a) and (b) above, explicates the claim that $S$ is synonymous with itself. Thus explained, the claim that a definitional abbreviation "creates" a new synonymy is entirely superficial and syntactical. What a definitional abbreviation does is provide us with a new way of rewriting, or transcribing, an already existing synonymy between a meaningful sentence $S$ and $S$ itself, where the 'synonymy' in question is explained in purely logical terms.

## 6.6   Synonymies without Explicit Definitions?

A definitional abbreviation $A_s$ of a sentence $S$ creates by stipulation an expression, $A_s$, for which one's language contains only one explanation, or criterion, for its use, namely, the explanation, or criterion, expressed by uses of *S*. Abbreviations of this sort are explicitly introduced words for which we have only one criterion of application. As Hilary Putnam points out in Putnam 1962b, p. 65, English contains some words that are similar to explicitly introduced one-criterion words, in the following sense: for each such word *w*, there is a single criterion *c* for the application of *w* such that in many contexts, and for most purposes, we apply *w* on the basis of *c* and we would explain the meaning of *w* by stating *c*. In the case of 'Bachelor', for instance, the criterion is 'unmarried man'. There are of course many problems with the criterion. Consider a man whose long-term companion and lover is someone he is not permitted to marry because gay marriage is illegal in his state or country. Is he a bachelor? Most would find it odd to say "Yes." By and large, however, and for most purposes (especially now that gay marriage is legal in many Western countries) we are content with the criterion, even if we don't like the word 'bachelor'. This is so not because we explicitly introduced the word 'Bachelor' as short for 'unmarried man', but simply because we typically use 'Bachelor', if we use it at all, as short for 'unmarried man'. Moreover, Quine agrees with Putnam that there are one-criterion words, including 'Bachelor' (Quine 1960, pp. 56–57), in English.

Here then is a grain of truth to the synonymy-by-usage criticism: the narrow notion of "synonymy by usage" that is clarified by cases of synonymy created by explicitly conventional definitions may be similar in varying degrees, and relative to particular explanatory purposes, to the relationships between the one-criterion words of natural languages, such as 'Bachelor', which are not introduced by explicit definition, on the one hand, and the longer phrases, such as 'unmarried man', that we use to state the criteria for applying those words, on the other.

This does not vindicate the synonymy-by-usage criticism, however. Both Quine and Putnam observed that many of our words occur in a vast number of theoretical sentences in such a way that no single explanation of the use of such words is suitable for all our theoretical purposes. Moreover, our explanations of the use of such words may vary widely without leading us to suspend or reject a homophonic translation between our past or present uses of the words and our current idiolect. Putnam calls these "law cluster" words (Putnam 1962b, pp. 50–54); Quine says they "connote a clusters of traits" (Quine 1960, p. 57). The problem for the synonymy-by-usage criticism is that we cannot explain how to apply the notion of synonymy to such words by looking for similarities with synonymies created by explicitly conventional definitions. Quine's doubts about the idea of synonymies that hold between two given expressions by dint of how they are used, independently of any explicitly conventional acts of definitional abbreviation, are primarily, even if not exclusively, doubts about the idea of synonymy for law cluster words.

Focusing on 'Bachelor', which is in some ways similar to an explicitly introduced definitional abbreviation, Grice and Strawson saw that if a transparent relation of synonymy can be created by an explicit definitional abbreviation, then there may be other, similar relations that were not so created. What they missed is that there is little or no similarity between most of the pairs of terms that are of interest in science, on the one hand, and pairs of terms related by explicitly conventional abbreviations, on the other. Their synonymy-by-usage criticism is therefore mistaken. There is no tension between Quine's claim that definitional abbreviations create synonymies and his skepticism about finding a general definition of synonymy.

CHAPTER 7

# Can First-Order Logical Truth Be Defined in Purely Extensional Terms?

I shall address the question whether first-order logical truth can be defined extensionally by reexamining P. F. Strawson's and W. V. Quine's seminal debate about it. I aim to show that a careful, sympathetic, and historically informed investigation of this largely forgotten debate clarifies the question and positions us to see how to construct and defend an affirmative answer to it.

## 7.1 Historical Background

Strawson's and Quine's debate about logical truth took place against the background of a broader clash between approaches to philosophy characteristic of Oxford philosophy in the 1950s, on the one hand, and Quine's systematic criticisms of the notions of synonymy and analyticity, on the other. Quine planted the seeds of his criticisms of these notions in his paper "Truth by Convention," first published in 1936. At Harvard in 1940–1941, in private discussions with Carnap, Quine and Alfred Tarski objected to Carnap's notion of analyticity, and proposed various ways of defining logical truth without relying on it (Mancosu 2005 and Frost-Arnold 2013). At roughly the same time, Quine was writing *Elementary Logic* (Quine 1941), his first logic text for undergraduates, and began working out the details of the purely extensional account of logical truth that he would later present in his textbook, *Methods of Logic*, which was first published in 1950, the same year in which Quine presented his revolutionary paper "Two Dogmas of Empiricism" at a meeting of the American Philosophical Association in Toronto. As Quine reports in Quine 1991,

> The response [to "Two Dogmas"] was quick and startling. The paper appeared in *Philosophical Review* a few weeks after the Toronto meeting, and four months later there were symposia on it in Boston and Stanford. (Quine 1991, p. 393)

144

It was at the height of this first wave of excitement about Quine's criticisms of meaning and analyticity that Quine was invited to be the George Eastman Professor at Oxford for the year 1953–1954. The philosophical scene as Oxford at that time was very different from the scene in the United States. Gilbert Ryle, the editor of *Mind* from 1948 to 1971 and an advocate of a version of Oxford ordinary language philosophy, was a dominant figure at Oxford, as was J. L. Austin, who, more than any other philosopher of the period, came to define and exemplify Oxford ordinary language philosophy. Strawson was of a younger generation than either Ryle or Austin, but had recently received a great deal of attention for his trenchant criticisms in "Truth" (Strawson 1950a) of Austin's account of truth, and, especially, for his revolutionary rejection, in "On Referring" (Strawson 1950b), of Bertrand Russell's analysis of definite descriptions.

The heart of Strawson's rejection of Russell's analysis of definite descriptions is Strawson's claim that

> 'mentioning', or 'referring', is not something that an expression does; it is something that someone can use an expression to do. Mentioning, or referring to, something is a characteristic of *a use* of an expression, just as 'being about' something, and truth-or-falsity, are characteristics of *a use* of a sentence. (Strawson 1950b, p. 326)

This attitude toward referring goes hand in hand with Strawson's observation, in his book *Introduction to Logical Theory* (Strawson 1952), that "we cannot identify that which is true or false (the statement) with the sentence used in making it, for the same sentence may be used to make quite different statements, some of them true and some of them false" (Strawson 1952, p. 4). These observations about referring and making statements led Strawson to regard the generalizations characteristic of deductive logic as true in virtue of the intensional properties of particular uses of linguistic expressions, especially those of ordinary language. One might hope to define logical truth without mentioning intensional properties of particular uses of linguistic expressions by introducing an artificial extensional language whose terms are supposed to be unambiguous. In Strawson's view, however, any linguistic expression, even a supposedly unambiguous expression of an artificial language, may be used in a variety of different ways. He therefore saw no way to define or explain logical truth in terms of a regimented language whose expressions we take to be unambiguous. He proposed, instead, that we define logical truth in terms of the intensional properties of *statements* expressed by our uses of sentences, whether they be ordinary language sentences, or sentences of an artificial language. In particular, in his view, a first-order logical truth is an *analytic* statement that exemplifies a first-order logical form all of whose instances are *analytic*.

The debate between Strawson and Quine got started when, just before Quine was to visit Oxford, Ryle invited him to review Strawson's *Introduction to Logical Theory*. Quine rarely accepted review assignments, but made an exception in this case, envisioning his review of Strawson's logic book "as a manifesto to start [his] Eastman year at Oxford" (Quine 1986a, p. 30). The completed review, which is indeed a manifesto, written in Quine's most brilliant style, was published as the lead article in *Mind*, October 1953. In it Quine forcefully argues, contra Strawson, that analyticity is "too soft and friable a keystone" for defining logical truth (Quine 1953a, p. 140) and recommends, instead, that we define a first-order logical truth as a *true* statement that exemplifies a first-order logical form all of whose instances are *true*. Given a fixed list of logical vocabulary and a Tarski-style definition of truth, Quine writes,

> the business of formal logic is describable as that of finding statement forms which are logical, in the sense of containing no constants beyond the logical vocabulary, and (extensionally) valid, in the sense that all statements exemplifying the form are true. Statements exemplifying such forms may be called *logically true*. Here there is no ... effort to separate the analytic from the synthetic. (Quine 1953d, pp. 140–141)

While Quine's extensional account of logical truth must have seemed completely unacceptable to Strawson for reasons I sketched above, Quine develops and defends it from a point of view that is so different from Strawson's that it would not have been easy for Strawson to find precisely the right point at which to challenge it. We do not know whether Strawson tried to respond to Quine's criticisms in seminars at Oxford in 1953–1954. According to Michael Dummett, however,

> Quine surely won most of the many jousts he had with Oxford philosophers [in 1953–1954]. Few of the latter had read much of Quine's work; most expected him to be easy prey to their refutations of him. I remember a meeting of the joint seminar run by Peter Strawson and Paul Grice at which Quine was present. Grice read a paper attacking Quine's views, which he finished with a triumphant glint in his eye. Quine replied to him with a subtle and crushing retort; Grice took on a very disconcerted expression. The only one who took the measure of Quine was John Austin, who read a paper to the Philosophical Society criticizing a minute point Quine had made in a footnote. (Dummett 2007, p. 51)

In any case, it was not until 1957, in his paper "Propositions, Concepts, and Logical Truths," that Strawson responded in print to Quine's criticisms, and even then he did so only indirectly, by criticizing Quine's proposed extensional alternative. Strawson 1957 argues that one cannot say what it is for a sentence or statement to exemplify a logical form without appealing to intensional notions, and hence that Quine's efforts to define first-order logical truth in purely extensional terms cannot succeed.

Quine did not immediately reply to this challenge (in Quine 1960, p. 65n, he describes it as "interesting," and says he cannot claim to have answered it anywhere). His only explicit attempt to reply to it is in Quine 1969d, where he characterizes what he takes to be the heart of the challenge, and offers an austere extensional account of logical truth that skirts the challenge by bypassing the difficult question of what it is for a statement to exemplify a logical form.

For reasons I shall explain below, however, Quine's 1969 reply to Strawson is uncharacteristically confused in ways that have not yet been noticed in the literature. This circumstance may at first seem to count in favor of Strawson's view. In fact, however, as I shall try to show, a proper analysis of the difficulties that Quine's reply faces suggests a new Quinean response to Strawson's challenge, one that supports Quine's view that logical truth can be defined in purely extensional terms.

The new Quinean response that I shall offer saves Quine's account from what may appear to be a devastating criticism, and shows that Quine's account is still worthy of serious consideration in contemporary philosophy of logic. The response that I shall offer does not address more recent criticisms of Quine's extensional account of logical truth, such as those raised in Etchmendy 1990 and 2008, or the related criticisms of Tarski's account of logical consequence – criticisms which are still much discussed today. Etchemendy's central criticism of Quine's account of logical truth is not that it cannot be stated in extensional terms, however, but that it does not capture our pretheoretical concept of logical truth. To address Etchemendy's criticisms and evaluate the recent literature about them, a defender of Quine's extensional account of logical truth must first address Strawson's more fundamental criticisms, as I try to do in this chapter.

## 7.2   Quine on Regimentation and Logical Truth

As Quine notes, on both his and Strawson's accounts of logical truth, "Logic ... is *formal* logic in a narrow sense which excludes those preparatory operations, in applied logic, whereby sentences of ordinary language are fitted to logical forms by interpretation and paraphrase" (Quine 1953d, p. 142; see also Quine 1940, p. 5). Quine's name for the process by which "sentences of ordinary language are fitted to logical forms by interpretation and paraphrase" is *regimentation*. Unlike Strawson 1957 (pp. 20–21), who believes that our regimentations should preserve the content of the particular utterances of ordinary language sentences that we aim to regiment, Quine 1960 (chapter 5) stresses that what counts as a suitable regimentation is up to the speaker who introduces it. While Quine's account of regimentation is partly motivated by his theoretical doubts about the objectivity of the relation of sameness of content, or synonymy, Quine argues that a requirement such as Strawson's, according to which a regimentation of an utterance $u$ should preserve the content of $u$, would be "out of place ... even if the notion of synonymy as such were in the best of shape" (Quine 1960, p. 208). "If we paraphrase a sentence to resolve ambiguity," for instance, he argues,

> what we seek is not a synonymous sentence, but one that is more informative by dint of resisting some alternative interpretations. Typically, indeed, the paraphrasing of a sentence $S$ of ordinary language and logical symbols will issue in substantial divergences. Often the result $S'$ will be less ambiguous than $S$, often will have truth values under circumstances under which $S$ has none..., and often it will even provide explicit references where $S$ uses indicator words. (Quine 1960, pp. 159–160)

Quine concludes that what counts as the "same" for the purposes of a given regimentation is up to the speaker who introduces the regimentation. When a speaker decides to use a sentence $S'$ of an artificial first-order language in place of a sentence $S$ of his ordinary language, the relation of $S'$ to $S$

> is just that the particular business that the speaker was on that occasion trying to get on with, with the help of $S$ among other things, can be managed well enough to suit him by using $S'$ instead of $S$. (Quine 1960: 160)

On this pragmatic account, a regimentation $S'$ of an ordinary language sentence $S$ need not and typically will not be synonymous with $S$.

Let us say that a *regimented language* is an artificial language whose sentences a person has decided to use for various purposes *in place of* his or her natural language sentences, and that a regimented language is *first-order* if it has a first-order logical grammar of the kind that is familiar from introductory logic books (such as Quine 1982). To simplify my exposition, I shall also suppose that a regimented language comprises all and only sentences constructed in the usual way from a finite list of basic predicates, such as 'lobster' and 'loves,' the truth-functional symbols '~', and 'V', identity sign '=', quantifier symbol '∀', and an unlimited stock of variables '$x_1$', '$x_2$', .... (proper names and functors can be paraphrased as descriptions in well-known ways that I shall not review here.)

Once we have adopted a first-order regimented language *RL*, it is natural to suppose that we may take a *second*, additional step, by using Tarski's method to define 'true-in-*RL*' in terms of satisfaction, where the satisfaction clauses for the basic predicates of the language, such as 'lobster' and 'loves', are disquotational. For it is natural to suppose that if *RL* is a language we can use, then we can apply Tarski's method to define, in a distinct metalanguage *MRL*, a predicate 'true-in-*RL*' that entails (nearly enough, with a qualification I shall note below) all and only the biconditionals of the form

(T) '_____' is true-in-*RL* if and only if _____,

where the blanks are uniformly filled by *RL* sentences.

The basic idea behind Tarski's method is that for each predicate of *RL*, one can adopt a satisfaction clause with a disquotational pattern, such as:

(Sat1) For every sequence *s*, *s* satisfies my word '_____' followed by var(*i*) if and only if $s_i$ is _____.

(Sat2) For every sequence *s*, *s* satisfies '_____' followed by var(*i*) and var(*j*) if and only if _____ $s_i$ $s_j$,

where $s_i$ and $s_j$ are the *i*th and *j*th objects in the sequence *s*. And so on. One can then complete one's definition of satisfaction for *RL* by adding the following clauses:

(Neg) For all sequences *s* and sentences *S*: *s* satisfies ⌜~*S*⌝ (the negation of *S*) if and only if *s* does not satisfy *S*.

(Disj) For all sequences *s* and sentences *S* and *S'*: *s* satisfies ⌜ *S* V *S'* ⌝ (the disjunction of *S* with *S'*) if and only if either *s* satisfies *S* or *s* satisfies *S'*.

(All) For all sequences *s*, sentences *S*, and numbers *i*: *s* satisfies ⌜ ∀*xᵢS* ⌝ (the universal quantification of *S* with respect to var(*i*)) if and only if every sequence *s'* that differs from *s* in at most the *i*th place satisfies *S*.

Together with the disquotational satisfaction clauses for the simple predicates of language *RL*, clauses (Neg), (Conj), and (All) define satisfaction for all sentences of *RL*. Using this definition of satisfaction, one can then define 'true-in-*RL*' as follows:

(Tr) A sentence $S$ of $L$ is true-in-$L$ if and only if $S$ is satisfied by all
sequences.[1]

For Quine a central appeal of defining true-in-$RL$ in the way I just
sketched is that to do so, one does not need to presuppose a fully general
and objective relation of translation; it is enough to specify a particular
relation that shall count as "translation" between sentences of $RL$ and
sentences of the metalanguage in which one specifies the relation. On
this approach, as Saul Kripke explains it when summarizing a similar
approach of Donald Davidson's, we "let the truth theory itself determine
the translation of the object language into the metalanguage" (Kripke
1976, p. 338). Just as we settle a suitable regimentation of an utterance $u$
by adopting a regimentation of $u$ that suits our goals for the regimentation, according to Quine, so we settle a "translation" of an object language into a metalanguage by adopting a particular truth theory that
suits our goals for the truth theory, which may not and typically will not
include the goal of being faithful to all aspects of our pre-theoretical concepts of translation and truth.

This view of translation in a theory of truth will seem wrong-headed to
those philosophers, such as Anil Gupta (in Gupta 1993), who think that
a central goal of a satisfactory truth theory for a particular language is to
be faithful to our pretheoretical concepts of translation and truth. While
such criticisms helpfully highlight the minimalist, pragmatic, and revi-
sionary aims of Quine's applications of Tarski's method of defining truth,
they do not pose an *internal* challenge to these applications, i.e., a challenge that can get a grip even if one is willing to let one's stipulations of
clauses for a Tarski-style truth theory for a given $RL$ settle what counts as a
"translation" of $RL$ into one's metalanguage. In contrast, as I shall explain
below, Strawson's criticism in Strawson 1957 does raise such an internal
challenge for Quine, a challenge that Quine himself acknowledges and
tries unsuccessfully to address.

When 'truth-in-$RL$' is defined in the minimalist, pragmatic way
sketched above, it is as clear to us as the satisfaction clauses of its inductive
definition. Since the satisfaction clauses for the basic predicates of $RL$ are
disquotational, these clauses are as clear to us as the basic predicates themselves. The satisfaction clauses for the logical constants are equally lucid,
despite not being purely disquotational. Since the satisfaction clauses for
the logical constants are not purely disquotational, the biconditionals

---

[1] These definitions apply methods that were discovered by Tarski and are systematically presented in
Tarski 1936. The particular formulations are based on Quine 1986d, chapter 3.

implied by a Tarski-style definition of 'true-in-*RL*' are not purely disquotational (this is the qualification I mentioned above), but they are just as clear to us as disquotational biconditionals of the form (T). In short, when 'truth-in-*RL*' is defined in this way for a regimented language *RL* that we can use, 'truth-in-*RL*' appears to be a clear and unproblematic predicate. (I shall challenge and qualify this claim below.)

According to Quine, *logical truth* for particular sentences of *RL* may now be defined in terms of 'true-in-*RL*'. We know that a given logical truth, *S*, of *RL* is true-in-*RL*. But what distinguishes *S*, as a logical truth, from other sentences that are true-in-*RL*? Quine's answer is that any sentence *S'* of *RL* that results from *S* when we uniformly substitute, for any sentence or predicate that occurs in *S*, some other sentence or predicate of *RL* is *also* true-in-*RL*. As Quine himself puts it, more loosely, "If we suppose a prior inventory of logical particles, comprising 'no', 'un-', 'not', 'if', 'then', 'and', etc., then in general a logical truth is a statement which is true and remains true under all reinterpretations of its components other than the logical particles" (Quine 1953b, pp. 22–23).

The same basic definition of logical truth can be formulated in terms of logical schemata. For instance, on Quine's view, to say that the *RL*-sentence '$\forall x_1((x_1 \text{ is a lobster}) \rightarrow (x_1 \text{ is a lobster}))$' is a logical truth is to say both that

(i) Every *RL* sentence of the form '$\forall x_1(Fx_1 \rightarrow Fx_1)$' is true-in-*RL*.

and

(ii) The *RL*-sentence '$\forall x_1((x_1 \text{ is a lobster}) \rightarrow (x_1 \text{ is a lobster}))$' can be obtained by uniformly substituting the *RL*-predicate 'is a lobster' for '*F*' in the schema '$\forall x_1(Fx_1 \rightarrow Fx_1)$'.

More generally, according to Quine, a first-order logical schema $\varphi$ is *logically valid* if and only if every *RL*-sentence that results when we substitute *RL*- sentences or predicates for sentence- or predicate-letters in $\varphi$ is true-in-*RL*, and an *RL*-sentence *S* is a *logical truth* if and only if *S* can be obtained by substituting *RL*- sentences or predicates for sentence- or predicate-letters in a valid first-order logical schema.

There is no guarantee that for every regimented language *RL*, if every *RL*-sentence that results by substitution from a logical schema $\varphi$ is true-in-*RL*, then $\varphi$ is logically valid in the standard model-theoretic sense – i.e., true for all interpretations in all nonempty domains of discourse. There is, however, a remarkable theorem in recursion theory – Quine calls it the *Löwenheim–Hilbert–Bernays (LHB)* theorem – which establishes that if one's first-order regimented language *RL* includes the identity sign and predicates that express addition and multiplication, then a first-order schema $\varphi$ is true for all interpretations in all nonempty domains of discourse if and only if every *RL*-sentence that can be obtained by substituting *RL*- sentences or predicates for sentence- or predicate-letters in $\varphi$ is true-in-*RL* (Quine 1982, pp. 211–212, and Kleene 1952, theorem 35; for an exposition of Kleene's complex proof of the LHB theorem, see Ebbs 2015). Quine concludes that for any regimented language *RL* that contains the identity sign and predicates that express addition and multiplication, we may use the above substitutional definition of logical truth for sentences of *RL* *in place of* the standard model-theoretic definition of logical truth.

Given a standard syntactical account of logical schemata and a suitably regimented language *RL*, the notion of substitution that is relevant to Quine's account of logical truth is purely syntactical (Quine 1982, chapters 26 and 28). And, as we saw above, the notion of truth-in-*RL* is as clear to us as the recursion clauses of its inductive definition. Hence the terms in which Quine defines logical truth, namely, 'logical schema', 'substitution', and 'truth-in-*RL*', are apparently clear and unproblematic.

In contrast, according to Quine, our application of 'true-in-*RL*' to an ordinary language sentence *S*, as uttered on occasion *O*, depends on our selection of a paraphrase, or translation, of *S*, as uttered on *O*, by some sentence *S′* of *RL*. (For a similar account of Quine's application of truth to sentences of ordinary language, see Harman 1967, pp. 150–151. I say more about this below.) We can classify an ordinary language sentence *S*, as uttered on *O*, as logically true or not in Quine's sense, only relative to a translation of *S* by some sentence *S′* of *RL*; given such a translation, *S* counts as logically true if and only if *S′* is logically true-in-*RL* in the substitutional sense defined above.

In short, on Quine's view, both truth and logical truth are directly defined only for particular regimented languages; truth and logical truth can be defined for particular uses of sentences of ordinary language, but only indirectly, relative to paraphrases of those uses of ordinary language sentences by sentences of a regimented language. He defends this division of labor in logical theory as follows:

> By developing our logical theory strictly for sentences in a convenient canonical [i.e., regimented] form we achieve the best division of labor: on the one hand there is theoretical deduction and on the other hand there is

the work of paraphrasing ordinary language into the theory. The latter job is the less tidy of the two, but still it will usually present little difficulty to one familiar with the canonical notation. For normally he himself is the one who has uttered, as part of some present job, the sentence of ordinary language concerned; and he can then judge outright whether his ends are served by the paraphrase. (Quine 1960, p. 159)

Quine's pragmatic, context-sensitive account of how truth and logical truth apply to particular uses of sentences of ordinary language is not (as Deckert 1973 and Boër 1977 believe) one of the central points at issue in the Strawson–Quine debate, but, for reasons I shall explain in more detail in §7.7 below, it is crucial to a full understanding of the Quinean account of logical truth that I shall propose.

## 7.3  Strawson's Criticism

Strawson's criticism in Strawson 1957 is aimed at the characterization of logical truth that Quine offers in "Two Dogmas of Empiricism" (Quine 1953b, pp. 22–23, quoted above), where Quine does not highlight the crucial step of regimentation. The criticism in Strawson 1957 may therefore seem irrelevant to Quine's more careful formulations of his account of logical truth in his well-known logic texts, including Quine 1940 and 1982, in which he pays careful attention to regimentation. Strawson's focus on the application of logic to ordinary language sentences may also suggest that his central criticism of Quine concerns the relationship between regimented language, for which Quine defines logical truth in the first instance, and ordinary language, to which Quine's account applies only via paraphrase, or translation, as described above. (Deckert 1973 and Boër 1977 interpret Strawson's criticism in this way.) As I shall reconstruct it here, however, Strawson's criticism strikes at the heart of Quine's account of logical truth by challenging Quine to justify his assumption that a given regimented language $RL$ in terms of which he proposes that we define 'logically true-in-$RL$' is unambiguous. On my reading, Strawson observes, in effect, that even a language that results from Quine's pragmatic method of regimentation may in fact contain ambiguous predicates.

I will start with an instructive but problematic formulation of Strawson's criticism, and then refine it. Suppose, for instance, that we decide to use a regimented form of the English word 'bank' in $RL$ without noticing its ambiguity. (Strawson's central example of an ambiguous word is 'sick', which can be used to mean, for instance, 'physically ill' or 'depressed'. See Strawson 1957, pp. 16, 19.) Now suppose that on a given occasion $O$, we use the $RL$ sentence

(a)  $\forall x_1((x_1 \text{ is a bank}) \rightarrow (x_1 \text{ is a bank}))$

in such a way that the first occurrence of 'bank' is satisfied by riverbanks only, and the second occurrence of 'bank' is satisfied by money banks only. Then it seems we should say that the resulting sentence of $RL$ is *not true*. But (a) is a sentence of the schematic form '$\forall x_1(Fx_1 \rightarrow Fx_1)$', and, according to the schematic conception of logical truth, every sentence of that schematic form is true-in-$RL$. If 'true-in-$RL$' is a good replacement for 'true', however, we should conclude that (a), as we used it on occasion $O$, is not true-in-$RL$.

This first formulation of Strawson's criticism is problematic, however, since if $RL$ is ambiguous (i.e., contains ambiguous predicates or logical constants), we can't define truth directly for $RL$ in Tarski's way. For a given *unambiguous* regimented language that we can directly use, call it $RL'$, we can define, in a distinct metalanguage $MRL'$, a predicate 'true-in-$RL'$' that entails all and only the biconditionals of the form

(T')  '_____' is true-in-$RL'$ if and only if _____,

where the blanks are uniformly filled by $RL'$ sentences. Tarski shows how to define such a predicate in terms of an inductive definition of satisfaction. If we put (a) in the blanks of its corresponding disquotational pattern, namely, (T), we get:

(t)  '$\forall x_1((\text{bank } x_1) \rightarrow (\text{bank } x_1))$' is true-in-$RL$ if and only if $\forall x_1((\text{bank } x_1) \rightarrow (\text{bank } x_1))$.

And, by assumption, this tells us nothing, and fails to settle how to apply 'true-in-$RL$' to (a), unless, somehow, the occurrences of 'bank' on the right-hand side of the biconditional are disambiguated. The problem is that if sentences of $RL$ are individuated by their spelling alone, and 'bank' is ambiguous in $RL$, we cannot suppose that occurrences of 'bank' on the right-hand side of the biconditional are disambiguated. We cannot make sense of using 'bank' without disambiguating it, and so we cannot make sense of affirming (t). In short, if $RL$ is ambiguous in the way we have assumed, we cannot then define 'true-in-$RL$' in Tarski's way.

The same problem shows up for particular satisfaction clauses of an inductive definition of truth for an $RL$ that contains the ambiguous word 'bank'. Using the pattern (Sat 1), one might try:

(sat 'bank') For all sequences *s*, *s* satisfies-in-*RL* 'bank $x_i$' if and only if $s_i$ is a bank.

But if all occurrences of 'bank' in (sat 'bank') are ambiguous, then we cannot make sense of affirming (sat 'bank'), since we cannot use 'bank' on the right-hand side.

We can make sense of affirming a clause like (sat 'bank') only if we thereby implicitly decide to use 'bank' in one of its several different ways on the right-hand side. Then by accepting (sat 'bank') as part of our definition of truth for *RL*, we will thereby stipulate that occurrences of the syntactic type 'bank' always make the same, disambiguated contribution to the derivation of Tarski-style biconditionals for sentences of a regimented language *RL'* in which they occur. Such a stipulation would entitle us to suppose there is no ambiguity in 'bank' in the language *RL'*, and thereby enable us to define 'true-in-*RL'*' in Tarski's way. (I expand on this idea below.) We could then apply this definition of 'true-in-*RL'*' to sentences of *RL*, as used on particular occasions, but only relative to a translation of those sentences, as used on these occasions, by sentences of *RL'*. We may translate a particular utterance of (a), for instance, by a sentence of *RL'*, namely, '$\forall x_1((x_1$ is a bank$_1) \rightarrow (x_1$ is a bank$_2))$', where the subscripts indicate different unambiguous predicates that we explain as meaning 'riverbank' and 'money bank', respectively. Clearly, when the predicates 'bank$_1$' and 'bank$_2$' are introduced as part of a first-order regimented language *RL'* by these stipulations, '$\forall x_1((x_1$ is a bank$_1) \rightarrow (x_1$ is a bank$_2))$' is not true-in-*RL'*.

As I shall reconstruct it, then, from a Quinean point of view, Strawson's observation, applied to sentence (a), is that

(1) The sentence (a) of *RL*, as used in the way sketched above, may be translated by '$\forall x_1((x_1$ is a bank$_1) \rightarrow (x_1$ is a bank$_2))$', which is not true-in-*RL'*.

and

(2) *Relative to this translation* we can classify (a), as used in the way sketched above, as not true-in-*RL'*.

Strawson's criticism is that any sentence of a regimented language that we introduce by Quine's pragmatic method of regimentation may be ambiguous in the way that this discussion of sentence (a) illustrates.

Thus clarified, the moral of Strawson's criticism is that we cannot define logical truth substitutionally for a given language *RL* unless its expressions, individuated syntactically, are not ambiguous. The problem for

Quine is that to rule out ambiguity – in particular, to rule out that on different occasions of use, particular occurrences of a given sentence type of a regimented language *RL* that we introduce via Quine's pragmatic method of regimentation may have different truth values – we apparently need to suppose that the *meanings* of the expressions of *RL* are constant across all the particular occasions on which speakers use them, contrary to Quine's claim that he can define logical truth without relying on such "soft and friable" notions as meaning or analyticity.

It is important not to conflate this problem with the superficially similar but much more general problem of understanding how Quine can make sense of the familiar fact that we can use our sentences to make assertions. Grice and Strawson 1956 raise the latter problem. They argue that if, as Quine's austere account of language use implies, synonymy is not an objective notion, then our sentences are "meaningless," and hence we cannot use them to make assertions (Grice and Strawson 1956, p. 146). If we cannot use our sentences to make assertions, then of course we cannot use our sentences to express any truths at all, including logical truths, and hence Quine's account of logical truth cannot succeed. As I reconstruct it here, however, Strawson's criticism of Quine's account of logical truth does not rest on the philosophically tendentious claim that Quine's austere account of language use implies we cannot use our sentences to make assertions. (I explain why this claim is tendentious, and, from Quine's point of view, completely misguided, in Chapter 5 of this volume.) On my reconstruction of Strawson's criticism, Strawson grants that Quine can make sense of the familiar fact that we can use our sentences to make assertions. The heart of the criticism is that even if we grant Quine this much, he is not entitled to the assumption that the sentences of languages regimented in the pragmatic way that he recommends are unambiguous. The problem is that there is apparently no way to guarantee that the sentences of languages regimented in the pragmatic way that Quine recommends are unambiguous without relying on assumptions about meaning that Quine officially rejects.

## 7.4 Quine's Concessions

In his 1969 reply to Strawson, Quine concedes both that a language introduced by his pragmatic method of regimentation may contain ambiguous predicates, and that one cannot define logical truth directly in terms of substitution for a language whose predicates are ambiguous. Quine acknowledges his plight in the following passage:

We might content ourselves with the definability of logical truth for language regimentations in which this difficulty does not arise: *univocal* regimentations, in which the extension of a term stays the same from one occurrence to another. But wait: how can I even state this univocality law, without intensionalism? (Quine 1969d, p. 323)

Quine recognizes that he needs a way of addressing Strawson's challenge *without* supposing that we can only speak of the extension of a predicate relative to a particular occurrence of it. For, as Strawson observes, once we attempt to fix extensions of words by reference to particular occurrences of them, we find ourselves asking what an occurrence of a word could be such that it fixes the extension of the word, and it seems we are compelled to start "speaking ... of a *word as used in a certain sense, with a certain meaning*" (Strawson 1957, p. 22). To clarify and defend his extensional account of logical truth, Quine needs a way of addressing Strawson's challenge without speaking in this intensionalist way.

Quine would be happy to settle the question of univocality, if he could, by a stipulation that certain sentences are to hold true. For to stipulate that a given sentence *S* is to hold true is not to intend to use *S* with some intensionally individuated *meaning* or *sense*, but to fix *S*'s "meaning" only in the weak extensional sense of settling *S*'s truth value (Quine 1936, pp. 89–90).[2] Despite the appeal for Quine of settling the problem by such a stipulation, however, in his reply to Strawson, Quine makes only one representative attempt to settle the question in this way, and explains why it fails.

> I could say simply and extensionally that '$(x)(Fx \equiv Fx)$' is to hold true for every one place general term in the '*F*' positions, and similarly for many-place general terms; but this is not enough, for it does not preclude shifts of extension in contexts of other forms than '$(x)(Fx \equiv Fx)$.' (Quine 1969d, p. 323)

Not seeing any fundamental barrier to a extensional stipulation that settles univocality, but also not seeing how to provide one, Quine concedes, at least for the sake of the argument, that no extensional stipulation will settle that our regimented language is univocal. He therefore tries a different approach to solving the problem, one that takes what he calls "a long way around" (Quine 1969d, p. 323).

---

[2] Contrary to what many Quine interpreters, including Isaacson 1992 and Becker 2012, believe, Quine never gave up the view that some sentences, though not all of logic, can be true by convention in the sense he explains in §II of Quine 1936. See, for instance, Quine 1986c, p. 206. I discuss this issue in more detail in Chapter 3 of this volume.

## 7.5   Quine's "Long Way Around"

Quine's proposed solution to the problem is as follows:

> Start out with one of the known complete proof procedures for logic – the logic, specifically, of truth functions, quantification, and identity. It can be fashioned to prove sentences directly – all the sentences that are instances of valid logical schemata – so we may omit any talk of schemata as intermediate devices. By just setting forth this general proof procedure we can define what it is for a sentence to be, as we may say, logically demonstrable. Thus far no talk of logical truth, nor validity, nor truth. Indeed some of the sentences that are logically demonstrable in this sense may be false, because (to speak crypto-intensionally) of changes in the extension of the term from one occurrence to another. But now we are in a position to state extensionally an adequate univocality condition: *a regimentation of our language is univocal if all the logically demonstrable sentences are true.* For a regimented language that is in this sense univocal, finally, logical demonstrability is logical truth. (Quine 1969d, pp. 323–324)

Quine takes himself in this passage to have defined what he calls "weakly univocal," and he proposes that we define logical truth substitutionally, and purely extensionally, for a regimented language *RL* that is weakly univocal.[3]

Just as one would expect, given his other writings, Quine says that "the general notion of truth thus presupposed" – the notion of truth presupposed in his account of 'weakly univocal' – "is meant in Tarski's way" (Quine 1969d, p. 324). But this raises a problem for Quine's account of 'weakly univocal'. For as we saw above, if *RL* is ambiguous, we can't define truth directly for *RL* in Tarski's way. When we clarified Strawson's observation, we supposed that we had discovered an ambiguity in *RL*, and for that reason constructed a truth theory for a new, disambiguated, language *RL'*, with two separate clauses for 'bank'. For the purposes of his "long way around," however, Quine cannot think of truth in this way, since his term 'weakly univocal' is supposed to be defined for *any* regimented language, even those that are not, in fact, weakly univocal, hence, even for regimented languages that contain ambiguous predicates, such as 'bank'.

One might think one could solve this problem for Quine's proposed definition of 'weakly univocal' by extending, via translation, a Tarski-style definition of truth for a disambiguated language such as *RL'* to particular

---

[3] Here we should add that the condition that *RL* is rich enough to express elementary arithmetic – recall the *LHB* theorem.

utterances of an ambiguous language, such as our imagined *RL*. One problem with this strategy is that it relies on finding translations of particular utterances of *RL* into sentences of *RL'*. Unless we have a general routine for finding such translations – a way of identifying different meanings of predicates by type and circumstance of use, perhaps – we could not generalize over every sentence of *RL* in this way. The most serious problem with the strategy, however, is that to implement it we would first have to identify a weakly univocal language, such as our imagined *RL'*, for which we can define truth in Tarski's way. The strategy therefore presupposes that we have *already* identified a univocal language, and hence does not yield a fully general criterion for univocality, as Quine's long way around is supposed to do.

We can express this problem in the form of a dilemma: Either we don't presuppose *RL* is free of ambiguity, in which case we have no reason to think we can define truth for it in Tarski's way, hence no reason to think we can define 'weakly univocal' for it; or we suppose it is free of ambiguity, in which case we don't need to define 'weakly univocal', but we are stuck, again, with Strawson's challenge of say how we are entitled to that supposition.

One might think Quine could avoid this dilemma by defining 'true' not for sentences as individuated orthographically, but only for particular *utterances* of sentences, with the assumption that the sentences are always disambiguated by features of the context of utterance. Quine does sometimes suggest that he accepts this way of thinking about truth. He writes, for instance, that

> what are best seen as primarily true or false are not sentences but events of utterance. If a man utters the words 'It is raining' in the rain, or the words 'I am hungry' while hungry, his verbal performance counts as true. Obviously one utterance of a sentence may be true and another utterance of the same sentence may be false. (Quine 1986f, p. 13)

Note, however, that Quine is speaking here of truth or falsity for *ordinary language* sentences. This discussion occurs in the first chapter of his book *Philosophy of Logic*; in later chapters he defines logical truth in terms of Tarski-style truth predicates defined for regimented languages. As I explained above, in the full development of his view, truth is defined directly only for a regimented language; it can then be *extended* via translation to events of utterance of sentences of an ordinary language. Hence in the full, official, story, as I understand it, truth predicates that apply primarily to events of utterance drop out, to be replaced by Tarski-style truth predicates that apply directly only to unambiguous sentences of a regimented language. It is nevertheless instructive to consider whether Quine can save his account of weakly univocal by appealing to the idea that truth applies to events of utterance. On this proposal, we may suppose, none of the items (namely, events of utterance) for which truth would be defined is ambiguous. (If some events of utterance are ambiguous, then the proposal is hopeless from the start.)

The fatal problem with this proposal is that our standard regimented languages contain infinitely many sentences. There will never be a time at which every sentence of such a language has been uttered or written down. We therefore cannot state the definition of 'weakly univocal' for our standard regimented languages in terms of a method of defining 'true' that applies only to events of utterance. Quine's official solution to this problem has long been to define truth in Tarski's way for sentences considered orthographically, as sequences of sets of written or uttered tokens of characters or phonemes (Quine 1960, p. 195), not for particular events of utterance.[4] I conclude that Quine's account of 'weakly univocal' cannot be stated in terms of a notion of truth that is directly defined only for events of utterance, and not for sentences individuated orthographically. I therefore see no way to save Quine's account of 'weakly univocal' from the dilemma summarized above. In short, Quine's "long way around" fails to define a general notion of 'weakly univocal', and therefore fails to address Strawson's challenge.[5]

## 7.6 A New Quinean Reply to Strawson

Quine's account of regimentation leaves open the question of the univocality of a given regimented language, and hence leaves his account of logical truth, according to which logical truth can only be defined for a univocal regimented language, vulnerable to Strawson's challenge. We could render Strawson's challenge irrelevant, however, if we could somehow introduce our regimented languages in a way that entitles us to suppose that they are univocal without relying on assumptions about meaning that Quine rejects.

For this purpose, I propose that we build a Tarski-style truth definition into our specification of the regimented language RL in terms of which

---

[4] Sayward 1975 makes essentially the same point, but does not raise the question of how it affects Quine's account of 'weakly univocal'.
[5] There is therefore little point in pursuing other problems for it, such as those raised in Gottlieb 1974.

we aim to define logical truth substitutionally. The possibility of doing so is already suggested in my Quinean reconstruction of Strawson's criticism. The key idea is to "let the truth theory itself determine the translation of the object language into the metalanguage" (Kripke 1976, p. 338). In particular, as I suggested above, by adopting a satisfaction clause for a given syntactically individuated predicate, such as 'bank', we may thereby directly stipulate that occurrences of that predicate always make the same, disambiguated contribution to the derivation of Tarski-style biconditionals for sentences of a regimented language in which they occur. We do not need to talk about sameness or difference of *meaning* to stipulate that the predicates and logical constants of a newly defined *RL* are to be *satisfactionally univocal*, in the sense that the satisfaction clauses for predicates and constants of *RL* always make the same contribution to the derivations of Tarski-style biconditionals for *RL* sentences in which those predicates or constants occur. If we stipulate a truth theory for *RL* when we introduce it, and someone asks us whether *RL* is univocal, we can show him our stipulations, and demonstrate, by appeal to the recursive structure of the resulting definition of satisfaction for *RL*, that each predicate or constant of the *RL* makes the same extensional contribution to sentences of *RL* in which it occurs. In particular, we can derive the extension of an *RL* predicate $P$ from the satisfaction clause we stipulate for $P$. If $F$ is a one-place *RL* predicate, for instance, then

The extension of $F$ = {$x$: $x$ satisfies-in-*RL* $F$} = {$x$: $x$ is $F$}.

The key point is that reference (satisfaction) can be explicated in such a way that expressions (typographically individuated, relative to a particular *RL*) have unambiguous satisfaction conditions – the very satisfaction conditions stipulated for them by the satisfaction clauses of the recursively structured Tarski-style truth theory that one provides when one adopts *RL*.

Recall that Quine has no principled objection to solving Strawson's problem by stipulation. Also, as Quine once observed, "The grammar that we logicians are tendentiously calling standard is a grammar designed with no other thought than to facilitate the tracing of truth conditions" (Quine 1986f, pp. 35–36). As far as I know, however, Quine never considered combining these two parts of his view. He therefore overlooked the possibility of avoiding Strawson's problem by stipulating a truth theory for one's regimented language at the moment one adopts it.

One might think my proposal that we build a Tarski-style truth definition into our specification of the regimented language *RL* in terms of which we aim to define logical truth substitutionally does not really save a Quinean from Strawson's criticism, since to stipulate a theory of truth for a given *RL* is in effect to *mean* or *intend* to use our words in the metalanguage in which we state the truth theory in a particular way. Recall, however, that we must not conflate Strawson's criticism of Quine's account of logical truth with the superficially similar but much more general problem of understanding how Quine can make sense of the familiar fact that we can use our sentences to make assertions. In the dialectical context relevant to assessing Strawson's criticism, Quine is entitled to suppose that he can use his sentences to make assertions, and hence, in particular, to stipulate a truth theory for a regimented language. The key point is that one need not presuppose any general notions of sameness and difference of meaning, beyond similarities that matter for an inquirer's particular purposes, in order to use one's words to make assertions, and hence, also, in order to state a truth theory for a particular *RL*. Quine gets himself into trouble on this point only because he grants, for the sake of his dispute with Strawson, at least, that the process of regimentation is distinct from that of stipulating a truth theory for the resulting regimented language.

The notion of stipulation is to be understood here in Quine's nonintensional way, according to which there is no guarantee that we will not someday revise or reject a stipulation we make when we introduce a regimented language. A given stipulation as to the satisfaction conditions of a predicate may turn out, in retrospect, to seem wrong or untranslatable, because of what we later come to regard as ambiguities in the words we used to express the stipulation. For reasons I explained above, however, we could make sense of this possibility only from the standpoint of a new language for which 'true' is defined in Tarski's way. We cannot prove that a given set of semantic stipulations that we adopt specifies a regimented language whose sentences we will never under any circumstances come to regard as ambiguous. This may appear to pose a problem for my proposed Quinean response to Strawson's criticism, but it does not. For on the Quinean Tarski-style disquotational method of recursively defining satisfaction and truth that I propose, we must judge truth and univocality from some standpoint or other, the best we can muster at the time. There is in general no theory-transcendent standpoint from which to ask whether a given sentence is true, and hence, in particular, there is no theory-transcendent standpoint from which to determine whether we are "absolutely" right or wrong to classify a given regimented language as satisfactionally univocal, or a given sentence as logically true. We have no choice but to rely on our best current semantic stipulations. We are therefore entitled to rely on these stipulations and to take the regimented language thus specified to be satisfactionally univocal.

## 7.7  Objections and Replies

The central question at issue in the Strawson–Quine debate is whether by using only extensional resources that Quine accepts, one can introduce a regimented language *RL* that one is entitled to regard as univocal, so that one can then define first-order logical validity and logical truth for sentences of *RL* substitutionally. One might grant that a Quinean can answer this central question in the way I just sketched, but still believe that one will have to appeal to intensional notions in order to raise and properly evaluate the question whether a given *RL* introduced in this way furthers an inquirer's purpose in adopting it.

One might think, for instance, that (a) the *only* purpose one could have in adopting an *RL* is to evaluate claims and arguments expressed in ordinary language, (b) a full account of how the *RL* furthers one's purposes must include an account of how the regimented sentences are related to claims and arguments expressed in ordinary language, and (c) a satisfactory account of how the regimented sentences are related to claims and arguments expressed in ordinary language relies on intensional notions of the sort that Quine claims to be able to do without.

This reasoning goes wrong in its first step, namely, (a). Quine's central purpose in adopting particular regimented languages is to further scientific inquiry. The pragmatism about regimentation that primarily interests him is the sort that he sees as integral to our construction and clarification of scientific theories, and, for that reason, is of a piece with his scientific naturalism – "the recognition that it is within science itself, and not in some prior philosophy, that reality is to be identified and described" (Quine 1981, p. 21). In his view, the adoption of an *RL* is of a piece with scientific theorizing, and therefore "not to be distinguished from a quest of ultimate categories, a limning of the most general traits of reality" (Quine 1960, p. 161).[6] From Quine's point of view, the only difference between theorizing in physics and in logic is that

> whereas we can expound physics in its full generality without semantic ascent [i.e., the use of a Tarski-style truth predicate], we can expound logic in a general way only by talking of forms of sentences [e.g., "Every *RL*

---

[6]  Quine's account of explication in §53 of Quine 1960 and his related rejection of the method of conceptual analysis are both expressions of his commitment to scientific naturalism.

sentence of the form 'p or not p' is true-in-*RL*"]. The generality wanted
in physics can be got by quantifying over non-linguistic objects, while the
dimension of generality wanted for logic runs crosswise to what can be got
by such quantification. It is a difference in shape of field and not in content. (Quine 1960, p. 274)

The deduction theorem for classical first-order logic establishes, in effect,
that a clarification of logical laws of the sort that Quine recommends also
yields a clarification of deductive implications, where both the laws and
the implications are expressed in one's regimented language. Both of these
central theoretical purposes are satisfied by the adoption of a regimented
language that facilitates our inquiries, *as conducted in that regimented language*. It is a separate question how well the regimented languages enable
us to evaluate claims and arguments expressed in ordinary language.

Quine still needs to address this question, however, since we may introduce a regimented language in order to evaluate claims and arguments
expressed in ordinary language. Suppose, for instance, that part of my
purpose in including the regimented predicates 'bank$_1$' and 'bank$_2$' in the
regimented language *RL'* that I partly specified in §7.2 above is to evaluate
an argument that I first expressed in ordinary language, using the ambiguous word 'bank', disambiguated by my use of it on a given occasion,
*O*. Suppose, also, that I decide on a particular way of paraphrasing my
utterances (on *O*) of ordinary language sentences that contain 'bank', into
sentences of *RL'*. One might think we are still able to ask questions such
as whether I intended the extension of one of the regimented predicates,
'bank$_1$' and 'bank$_2$', to include the extension or intension of one of more
of my utterances (on *O*) of the ordinary language expression 'bank'. And
once this question is asked, it may seem that a Quinean is in an awkward
position. For, as I noted above, Strawson in effect argues (Strawson 1957,
p. 22) that

> (P) To speak of the *extension* of the ordinary word 'bank', as used on *O*,
> is to presuppose that there is a fact of the matter about the meaning
> or intension of 'bank' as used on *O* – a fact about meaning or inten-
> sion that, together with facts about the world, settles the extension of
> 'bank', as used on *O*.

If (P) is true, then there is a substantive, factual question of whether our
paraphrase or translation of 'bank', as used on *O*, by some predicate of our
regimented language fulfills our intention to include or be faithful to the
intension, and hence also the extension, of 'bank' as used on *O*. In this
way, according to the objection, when we apply logic to ordinary language

sentences we presuppose there are facts about the intensions expressed by our utterances of ordinary language words, contrary to Quine's view that we can do logic without presupposing that there are such facts.

As a first step toward answering this objection, recall that on Quine's account, both truth and logical truth are directly defined only for particular regimented languages. To this we can now add that logical truth is only defined for a regimented language that one is entitled to regard as univocal, for the reasons explained in §7.6. Truth and logical truth can be defined for particular uses of sentences of ordinary language, but only indirectly, relative to paraphrases of those uses of ordinary language sentences by sentences of a regimented language. In Quine's view, to speak of the *extension* of the ordinary word 'bank', as used on *O*, is in effect to choose a paraphrase, or translation, of the ordinary word 'bank', as used on *O*, into one's current regimented language. To apply logic to sentences we have used in the past, it is enough to choose such a paraphrase or translation into one's current regimented language – there is no need, in addition, to accept (P).

The second and final step in answering the objection is to note that if our purpose is to evaluate an utterance of an ordinary language sentence in the extensional terms available in first-order logic, then we need only choose some paraphrase of the utterance that suits our purposes in the context of evaluation. One might think that if one's purposes are, for instance, to disambiguate, then one will have to rely on intensional notions in order to choose a paraphrase. As I noted above, however, choosing a paraphrase is not typically a matter of finding a synonymous expression, and our choice of a paraphrase need not presuppose that synonymy is an objective relation. As Gilbert Harman explains:

> According to [Quine] a person associates varying paraphrases with ambiguous sentences. Paraphrases will be sentences in the speaker's language, not sentences or sentence substitutes in some universal language of propositional attitudes. The varying paraphrases represent what the speaker takes to be equivalent to the ambiguous sentence given a particular context. The notion of equivalence here is not that of "meaning equivalence," but rather the notion of an equivalence taken to follow fairly obviously from what the relevant group of people (possibly only the speaker or hearer) accepts. (Harman 1967, pp. 150–151)

The key point is that to use one's regimented language to evaluate a claim or argument expressed in ordinary language, it is enough to choose a paraphrase that provides an "equivalent" of the claims or the argument, where "equivalent" is understood in terms of what we and the speaker

in question (who may be our past self) accept. This is not to rely on the
assumption that there is an objective relation of synonymy; nor is it to rely
on a grasp of the intensions, if any, of the claims or arguments that one
paraphrases, in any sense of "intension" that Strawson takes for granted in
his criticisms of Quine.

The conclusion of the reasoning in the previous two paragraphs is that
one need not accept (P) in order to apply logic to evaluate one's ordinary language utterances. In addition, Quine is independently committed
by his empirical theory of translation (Quine 1960, chapter 2) to *rejecting* (P). His indeterminacy thesis implies that there is always more than
one acceptable paraphrase of an ordinary language sentence or term, and,
in general, no fact of the matter about which of the acceptable translations of sentences or of terms is correct. Quine's argument in Quine 1960,
chapter 2, also provides a framework for thinking about paraphrase, or
translation, in a way that accommodates our general deference (Quine
1960: 160) to a speaker's own choices about how to regiment her sentences
in a way that suits her purposes. When such choices are viewed in the context of Quine's indeterminacy thesis and his sharp theoretical distinction
between direct definitions of truth and logical truth, on the one hand,
and applications of such definitions to ordinary language utterances, on
the other, it becomes clear that Quine has principled reasons to reject (P).

To accept the Quinean account of logical truth that I recommend,
however, one need not be committed to the indeterminacy thesis. As
I explained two paragraphs above, what is crucial for our evaluation of
claims and arguments expressed in ordinary language is just that one's
"translations" from utterances of ordinary language sentences to sentences
of one's regimented language vary from context to context in ways that
reflect our decisions about which of the many sentences that the person
whose utterances we are translating accepts should come out true when
translated into our regimented language, given our present theoretical
purposes. To make such decisions is not to rely on a grasp of the intensions, if any, of the claims or arguments that one paraphrases, in any sense
of "intension" that Strawson takes for granted in his criticisms of Quine.

It should now be clear that on the Quinean view that I have sketched,
all of one's current judgments about satisfaction and truth are made from
the standpoint of a particular regimented language that one has adopted
via stipulations of satisfaction clauses for it, in the way described above.
These judgments may be licensed directly, by the definitions of satisfaction and truth built into our regimented language, or indirectly, by those
definitions together with translations of utterances of other sentences

into our regimented language. There is no guarantee that in the course of our theorizing, we will not later come to adopt a new regimented language and to judge, in retrospect, that a satisfaction clause we previously accepted is false or ambiguous. It is equally important, however, that there is no semantic exile – no theory-transcendent standpoint from which to determine if we are "absolutely" right or wrong to classify a given regimented language as satisfactionally univocal, or a given sentence as logically true. As I emphasized above, we have no choice but to rely on our best current semantic stipulations. We are therefore entitled to rely on these stipulations and to take the regimented language thus specified to be satisfactionally univocal.

I conclude that if we stipulate a Tarski-style truth theory for a regimented language *RL* that we propose to use, we are entitled to suppose that *RL* is satisfactionally univocal. If *RL* contains the identity sign and predicates that express addition and multiplication (recall the *LHB* theorem, again), we may therefore define first-order logical validity and logical truth in purely extensional terms, as Quine proposes: a first-order logical schema φ is *logically valid* if and only if every *RL*-sentence that results when we substitute *RL*-sentences or predicates for sentence- or predicate-letters in φ is true-in-*RL*, and an *RL*-sentence *S* is a *logical truth* if and only if *S* can be obtained by substituting *RL*- sentences or predicates for sentence- letters or predicate-letters in a valid first-order logical schema.

CHAPTER 8

# Reading Quine's Claim That No Statement Is Immune to Revision

Most critics and defenders of Quine's arguments in "Two Dogmas of Empiricism" (Quine 1953b) have read his claim that "no statement is immune to revision" as the claim that for *every statement S that we now accept there is a possible rational change in beliefs that would lead one to reject S*. For reasons I will explain below, I paraphrase this latter claim as follows:

> (R) For every sentence $S$ that a subject $A$ accepts at a time $t_1$, there is a possible rational revision of the beliefs $A$ holds at $t_1$ that (i) leads $A$, or another subject $B$, rationally to judge, at some later time $t_2$, that $S$ is false, and (ii) allows for a homophonic translation of $S$, as $A$ uses it at $t_1$, by $S$, as $A$ or $B$ uses it at $t_2$.

This standard reading of Quine's claim that "no statement is immune to revision" faces two serious problems. First, in "Two Dogmas" Quine does not even mention the issues about diachronic changes in belief or homophonic translation that are relevant to supporting (R). He writes as if he thinks he has a simple, direct argument that does not rest on such considerations. Second, Quine's own views about translation lead him to conclude, apparently contrary to (R), that some revisions in our current beliefs *would* alter the meanings of some of our words to the point that we would not accept a homophonic translation of some of the sentences we held true before the revision.

Until recently I accepted the standard reading despite these problems. On the reading I now prefer, however, Quine's claim is that no statement is immune to retraction,[1] understood as follows:

> (P) No statement we now accept is guaranteed to be part of every scientific theory that we will later come to accept.

---

[1] Quine uses the word "retract" (and "rescind") when discussing holism in Quine 1992: 14–15.

I will argue that in paragraph 2 of §6 Quine combines this uncontroversial observation with arguments and proposals from previous parts of "Two Dogmas" to support his conclusion that "it [is] folly to seek a boundary between synthetic statements, which hold contingently on experience, and analytic statements, which hold come what may" (Quine 1953b, p. 43). The key to my alternative interpretation is to see that in paragraph 1 of §6 Quine sketches a bold new naturalistic explication of the traditional notion of empirical confirmation, and that his aim in paragraph 2 is to show that the explication of confirmation he sketches in paragraph 1 is of no help in characterizing a boundary between analytic and synthetic statements. This is the last of a series of clarifications and observations in "Two Dogmas" that in Quine's view together show that it is folly to seek a boundary between analytic and synthetic statements, or, in other words, that we have no reason to suppose that such a distinction can be drawn in a language suited for and used in the mature natural sciences.

## 8.1    A First Look at the Context of Quine's Claim That "No Statement Is Immune to Revision"

Quine's claim that "no statement is immune to revision" appears in paragraph 2 of §6 (the final section) of "Two Dogmas of Empiricism." By the start of §6, Quine takes himself to have shown in the previous five sections of the paper that there is no way to draw an analytic-synthetic boundary in terms of an extensional specification of logical truth supplemented by definitions (§§1–2), an extensional specification of logical truth supplemented by a synonymy relation defined in terms of substitutivity *salve veritate* (§3), semantical rules laid down for an artificial language (§4), or confirmation by experience (§5).

In the first paragraph of §6, Quine likens science to "a field of force whose boundary conditions are experience," and sketches some consequences of this comparison, including the consequence that "no particular experiences are linked with any particular statements in the interior of the field, except indirectly through considerations of equilibrium affecting the field as a whole." In paragraph 2 he writes:

> If this view is right, it is misleading to speak of the empirical content of an individual statement – especially if it is a statement at all remote from the experiential periphery of the field. Furthermore, it becomes folly to seek a boundary between synthetic statements, which hold contingently on experience, and analytic statements, which hold come what may. Any statement can be held true come what may, if we make drastic enough

adjustments elsewhere in the system. Even a statement very close to the periphery can be held true in the face of recalcitrant experience by pleading hallucination or by amending certain statements of the kind called logical laws. Conversely, by the same token, no statement is immune to revision. Revision even of the logical law of the excluded middle has been proposed as a means of simplifying quantum mechanics; and what difference is there in principle between such a shift and the shift whereby Kepler superseded Ptolemy, or Einstein Newton, or Darwin Aristotle? (Quine 1953b, p. 43)

The paragraph begins with a conditional whose antecedent is "this view is right" and whose consequent is a conjunction of "it is misleading to speak of the empirical content of an individual statement" and – as signaled by the word "furthermore" – "it becomes folly to seek a boundary between synthetic statements, which hold contingently on experience, and analytic statements, which hold come what may." In the rest of the paragraph, Quine presents an argument in support of the conditional "if this view is right, then it becomes folly to seek a boundary between synthetic statements, which hold contingently on experience, and analytic statements, which hold come what may." The main premises of Quine's argument for this conditional are "any statement can be held true come what may" and "no statement is immune to revision."

## 8.2 The Standard Interpretation

In their influential paper "In Defense of a Dogma," H. P. Grice and P. F. Strawson say that Quine takes his assertion that no statement is immune to revision "to be incompatible with acceptance of the distinction between analytic and synthetic statements" (Grice and Strawson 1956, p. 154). They try to discredit Quine's argument in paragraph 2 of §6 by pointing out that there is a perfectly ordinary interpretation of "no statement is immune to revision" on which the statement is "not incompatible with acceptance of the distinction, but is, on the contrary, most intelligibly interpreted in a way quite consistent with it" (Grice and Strawson 1956, p. 154). They reason as follows:

> Any form of words at one time held to express something true may, no doubt, at another time, come to be held to express something false.... Where such a shift in the sense of the words is a necessary condition of the change in truth-value, then the adherent of the distinction will say that the form of words in question changes from expressing an analytic statement to expressing a synthetic statement.... If we can make sense of the idea that the same form of words, taken in one way (or bearing one sense), may express something true, and taken another way (or bearing another sense),

may express something false, then we can make sense of the idea of conceptual revision. And if we can make sense of this idea, then we can perfectly well preserve the distinction between the analytic and the synthetic,
while conceding to Quine the revisability-in-principle of everything we say.
(Grice and Strawson 1956: 157)

This challenge to Quine's reasoning has put Quineans on the defensive. In
response to it, Hilary Putnam (in Putnam 1962b and many other papers,
including Putnam 1979) argued, in effect, that even if we cannot *now* see
how we could judge that a sentence *S* is false without changing its meaning to the point where we would no longer interpret it homophonically
into our new body of beliefs, we may find *later* that we *can* judge that *S* is
false while still translating our previous uses of *S* into our (revised) theory
homophonically.[2]

Putnam's influential response to the Grice–Strawson criticism led many
interpreters (including me) to read Quine's claim that "no statement is
immune to revision" as the claim that no statement is immune to *rejection*, where the notion of rejection is understood partly in terms of homophonic translation. As Gilbert Harman explains, on Quine's view,

> there is no sharp, principled distinction between changing what one means
> and changing what one believes. We can, to be sure, consider how to trans-
> late between someone's language before and after a given change in view. If
> the best translation is the homophonic translation, we say there has been a
> change in doctrine; if some other (non-homophonic) translation is better,
> we say there has been a change in meaning. (Harman 1994, p. 141)

These Quinean points about meaning and homophonic translation are
integral to the standard interpretation of Quine's claim that no statement is immune to revision. According to Cory Juhl and Eric Loomis, for
instance,

> Quineans … introduce a surrogate for "means the same," when they
> appeal to the notion of a "good" or "best" translation scheme. A sentence
> "retains its meaning" across changes, on this Quinean picture, just in case
> the sentence would or should be translated homophonically across the
> change in language. But given this "surrogate" for synonymy, it seems as if
> Quineans can now make sense of what Carnap, Grice and Strawson, and a
> host of others are worried about when considering the possibility of "giv-
> ing up" a statement on the basis of empirical evidence. *It is not enough for*

---

[2] I say "in effect" because Putnam does not use the notions of rational revision or homophonic translation in Putnam 1962b. He did nevertheless implicitly defend something like (R) in that paper, and
in many subsequent ones. He explicitly talks about rational revision and homophonic translation in
Putnam 1979.

*the Quinean to show that we could give up our practice of asserting sentence*
*s, for some purportedly analytic s.* Rather, in order to address the worries of
his opponents, the Quinean must show that the sentence can be given up
while retaining its meaning across the change in language, that is, he must
show that the sentence is such that it could be translated homophonically across the change. *It may be that some sentences might plausibly stop*
*being asserted, but homophonic translatability imposes a further constraint,*
*and narrows the range of sentences which meet it. Whether any of the usual*
*examples (bachelorhood in the face of new marriage laws, etc.) meet this con-*
*straint is likely to remain controversial.* (Juhl and Loomis 2010, p. 116; my
emphases)

On this standard interpretation, to discredit the assumption that there is a
boundary between analytic and synthetic statements, as Juhl and Loomis
say, "it is not enough for the Quinean to show that we could give up our
practice of asserting sentence *s*, for some purportedly analytic *s*. Rather,
in order to address the worries of his opponents, the Quinean must show
that the sentence can be given up while retaining its meaning across the
change in language." On this interpretation, if Quine's claim that "no
statement is immune to revision" is to be understood in a way that engages
with the worries of his opponents, it must be understood as follows:

> (R) For every sentence *S* that a subject *A* accepts at a time $t_1$, there is a
> possible rational revision of the beliefs *A* holds at $t_1$ that (i) leads *A*, or
> another subject *B*, rationally to judge, at some later time $t_2$, that *S* is
> false, and (ii) allows for a homophonic translation of *S*, as *A* uses it at
> $t_1$, by *S*, as *A* or *B* uses it at $t_2$.

Following Grice and Strawson, critics of Quine's reasoning in paragraph
2 of §6 of "Two Dogmas" say, in effect, that in this paragraph Quine does
not make a convincing (or indeed, any) case for (R). Defenders of Quine's
reasoning in paragraph 2 have for the most part (tacitly) accepted this
criticism. Following Hilary Putnam, Quineans typically respond to the
criticism by trying to provide support for (R). It is now common for both
Quineans and critics of Quine to read (R) back into paragraph 2 of §6,
and to conclude that the argument there is at best incomplete.

## 8.3   Two Problems for the Standard Reading

There are two serious problems for the standard reading of Quine's reasoning in paragraph 2 of §6 of "Two Dogmas."

First, Quine's brief and sketchy presentation of his reasoning strongly
suggests that he thought a reader who adopts his "field of force"

description of science would not need much detail to be convinced of his claim that "no statement is immune to revision" and to agree that it implies that there are no analytic statements. If the claim amounts to (R), however, it is neither obvious nor uncontroversial. To support (R) one would need to show that radical changes in belief never bring about radical changes in meaning of the sort that would lead one to reject a homophonic translation of a previously uttered sentence into one's new theory. But Quine does not say anything about these difficult issues in the second paragraph of §6 (or anywhere else in "Two Dogmas"). He writes as if he thinks he has a simple, direct argument that does not rest on such considerations.

Second, in the Introduction to the first edition of *Methods of Logic*, published in 1950 – the same year in which Quine first presented "Two Dogmas of Empiricism" at the eastern meeting of the APA in Toronto and the year *before* that paper was published in the *Philosophical Review* – Quine writes that mathematics and logic are "central to our conceptual scheme" in the sense that they "can easily be held immune to revision on principle," and therefore "tend to be accorded such immunity, in view of our conservative preference for revisions which disturb the system least" (Quine 1950, p. xiii). He adds that "it is perhaps the same to say, as one often does, that the laws of mathematics and logic are true simply by virtue of our conceptual scheme" (Quine 1950, p. xiv), and then notes that

> it is also often said that the laws of mathematics and logic are true by virtue of the meanings of the words '+', '=', 'if', 'and', etc., which they contain. This I can also accept, for I expect it differs only in wording from saying that the laws are true by virtue of our conceptual scheme. (Quine 1950, p. xiv)

He also observes that

> mathematical and logical laws themselves are not immune to revision if it is found that essential simplifications of our whole conceptual scheme will ensue. There have been suggestions, stimulated largely by quandaries of modern physics, that we revise the true-false dichotomy of current logic in favor of some sort of tri- or *n*-chotomy. (Quine 1950, p. xiv)

The point Quine makes in the second sentence of this passage is clearly similar to his observation in the second paragraph of §6 of "Two Dogmas of Empiricism" that "revision even of the logical law of the excluded middle has been proposed as a means of simplifying quantum mechanics." But the crucial passage from Quine's Introduction to the first edition of

*Methods of Logic* – the passage that clearly conflicts with attributing (R) to Quine – is the following:

> Thus the laws of mathematics and logic may, despite all "necessity," be abrogated. But this is not to deny that such laws are true by virtue of the conceptual scheme, or by virtue of meanings. *Because these laws are so central, any revision of them is felt to be the adoption of a new conceptual scheme, the imposition of new meanings on old words.* (Quine 1950, p. xiv, my emphasis)

One might suspect that these passages were written before Quine finally made up his mind, while writing "Two Dogmas of Empiricism," to give up the analytic–synthetic distinction, and that Quine just did not get around to revising the passages before the first edition of *Methods of Logic* went to press. In fact, however, the crucial italicized sentence in the passage just quoted also appears in all *subsequent* editions of *Methods of Logic* – it appears on p. xiv of the 1959 revised edition (Quine 1959), and on p. 3 of both the third (Quine 1972) and fourth (Quine 1982) editions. Moreover, in "Carnap and Logical Truth," written in 1954, three years after "Two Dogmas" was first published, Quine argues that we should question our translation of an apparently sincere utterance if under the translation, the utterance counts as expressing the negation of an obvious logical law. As Quine later stressed, his view of translation, as applied to logical statements of a person's theory, implies that a logician who wishes to revise an established logical law faces a "predicament: when he tries to deny the doctrine, he only changes the subject" (Quine 1970, p. 81). As the above passages from the introduction to the 1950 edition of *Methods of Logic* show, this was clearly his view already at the time he wrote "Two Dogmas of Empiricism." He was therefore not in a position to affirm (R) for any sentence that he used to express one of these obvious logical laws. Yet, as we have seen, in paragraph 2 of §6, in support of his claim that no statement is immune from revision, he writes, "Revision even of the logical law of the excluded middle has been proposed as a means of simplifying quantum mechanics" (Quine 1953b, p. 43). This strongly suggests that Quine's claim that no statement is immune to revision should not be interpreted as (R).

Despite these problems, until recently I accepted the standard interpretation of Quine's reasoning in paragraph 2 of §6. Like many others, I assumed that Putnam-style counterexamples to analyticity – counterexamples that support, even if they do not conclusively establish, (R) – provide the best grounds for Quine's claim no statement is immune to revision. Thus bewitched, I did not look carefully for a plausible alternative to the standard interpretation, and, of course, did not find one, either. While recently trying to explain to myself (and my students) exactly what is going on in paragraph 2 of §6, however, I studied that paragraph more closely, and discovered a better interpretation, one that avoids the problems just explained.

## 8.4 An Alternative Interpretation

On the alternative interpretation that I now prefer, Quine's reasoning in paragraph 2 of §6 is the last of a series of clarifications and observations in "Two Dogmas" that in Quine's view together show that it is fruitless to seek a boundary between analytic and synthetic statements. Quine does not claim that there is no such boundary. His conclusion in paragraph 2 is that "it becomes folly to seek a boundary between synthetic statements, which hold contingently on experience, and analytic statements, which hold come what may." He is similarly subtle in §4, where he announces that

> for all its a priori reasonableness, a boundary between analytic and synthetic statements simply has not been drawn. That there is such a distinction to be drawn at all is an unempirical dogma of empiricists, a metaphysical article of faith. (Quine 1953b, p. 37)

These words signal that Quine is employing Carnap's own preferred method for revealing the emptiness of a philosopher's words – a method partly inspired by Wittgenstein's pronouncement, in his *Tractatus Logico-Philosophicus*, that

> the correct method in philosophy [is] to say nothing except what can be said, i.e. propositions of natural science – i.e. something that has nothing to do with philosophy – and then, whenever someone else wanted to say something metaphysical, to demonstrate to him that he had failed to give a meaning to certain signs in his propositions. (Wittgenstein 1921, pp. 73–74)

As I read it, Carnap's and Quine's dispute about the analytic–synthetic distinction is at root a dispute about whether such a distinction can be drawn in a language suited for and used in the mature natural sciences.

Quine's strategy in "Two Dogmas" is to demonstrate to Carnap and others that none of the minimally plausible strategies for giving clear meanings to the terms "analytic" and "synthetic" is successful, so the terms should be dropped from a properly scientific philosophy. To pursue this strategy, at each new step in his reasoning, including the step he takes in paragraph 2 of §6, Quine needs to rely on clarifications and observations that he makes in earlier parts of the paper. As signaled by its first sentence, which begins with the words "If this view is right," Quine's reasoning in paragraph 2 relies immediately and obviously on the view of science that he briefly sketches in the preceding paragraph, which I quote here in full:

> The totality of our so-called knowledge or beliefs, from the most casual matters of geography and history to the profoundest laws of atomic physics or even of pure mathematics and logic, is a man-made fabric which impinges on experience only along the edges. Or, to change the figure, total science is like a field of force whose boundary conditions are experience. A conflict with experience at the periphery occasions readjustments in the interior of the field. Truth values have to be redistributed over some of our statements. Reevaluation of some statements entails reevaluation of others, because of their logical interconnections – the logical laws being in turn simply certain further statements of the system, certain further elements of the field. Having reevaluated one statement we must reevaluate some others, which may be statements logically connected with the first or maybe the statements of logical connections themselves. But the total field is so underdetermined by its boundary conditions, experience, that there is much latitude of choice as to what statements to reevaluate in the light of any single contrary experience. No particular experiences are linked with any particular statements in the interior of the field, except indirectly through considerations of equilibrium affecting the field as a whole. (Quine 1953b, pp. 42–43)

In this paragraph Quine writes of "readjustments" occasioned by "a conflict with experience," readjustments that reflect "reevaluations of some statements." Since "there is much latitude of choice as to what statements to reevaluate in the light of any single contrary experience," what holds a statement $S$ in place in the "interior" of the field is not our recognition of a theory-independent standard of confirmation for $S$, but our actual, current *acceptance* of $S$ – our evaluation of $S$ as true – and the logical and explanatory relations between $S$ and other statements, relations that are themselves settled by our acceptance of statements of laws of logic, mathematics, and physics. Particular experiences are "linked with," and in that sense, *confirm*,[3] particular statements in the interior of the field only "indirectly through considerations of equilibrium affecting the field as a whole." In short, on the view Quine sketches in the first paragraph of §6,

---

[3] Although Quine does not use the word "confirmation" in the first paragraph of §6 of "Two Dogmas," he does use the word in the second paragraph, where he presents himself as drawing consequences from the view sketched in the first paragraph. He also uses the word "confirmation" in Quine 1960, p. 63–64, in the chapter in which he takes himself to be developing the account of confirmation sketched in "Two Dogmas".

an inquirer's acceptance of a statement establishes a relation between sentences and experience that does not exist apart from his or her acceptance of the theory, and to say that a statement is *confirmed* is just to say that one accepts it as part of one's best current theory.

This is a naturalistic explication (or appropriation, if you prefer) of Carnap's radical view that there is no relation of confirmation apart from our *decisions* about how to relate statements to experience. Quine's bold step is to do without Carnap's idea that some statements of a theory are analytic (i.e., true in virtue of semantical rules that we have laid down for the language of the theory) and others are synthetic (i.e., have truth values that are not settled by semantical rules that we have laid down for the language of the theory). For reasons that Quine explains in §4 of "Two Dogmas," he thinks Carnap fails to provide a satisfactory explication of this idea. Since all of Carnap's attempts to clarify the idea that some statements have empirical content presuppose his explications of analyticity in terms of semantical rules, once Quine takes himself to have shown that the idea of analyticity cannot be explained in terms of semantical rules, he can easily show in §5, drawing on observations about the holism of theory testing that Carnap himself had made in previous publications, that a boundary between analytic and synthetic statements cannot be drawn on the basis of relations between sentences and experiences that supposedly confirm them. (Many readers of "Two Dogmas" do not realize that Quine's point in §5 is one that Carnap himself would readily accept, namely, that if there is no independent account of the boundary between analytic and synthetic statements, an appeal to relations between sentences and experiences that supposedly confirm them will not help us to draw it.) What is left, according to Quine, as he later explained in *Word and Object*, §6, is just "scientific method" which "produces theory whose connection with all possible surface irritation consists solely in scientific method itself, unsupported by ulterior controls" (Quine 1960, p. 23).

In paragraph 2 Quine explains why one cannot exploit this minimalist explication of confirmation – the only positive bit of theorizing about confirmation that he presents in "Two Dogmas" – to draw a boundary between analytic and synthetic statements. He reasons as follows. If one grants his first point in the paragraph, namely, that one consequence of his minimalist explication of confirmation is that "it is misleading to speak of the empirical content of an individual statement – especially if it is a statement at all remote from the experiential periphery of the field," then the most plausible way to exploit his minimalist explication of confirmation to draw a boundary between analytic and synthetic statements, would be to equate "*S* is analytic" with a *new* explication of "*S* is confirmed come what may" – the same phrase for which in §5 of "Two Dogmas" he could find no clarification in terms of a supposedly substantive, theory-independent relation of confirmation by experience. In the context of the minimalist explication of confirmation that Quine sketches in the first paragraph of §6, to say that a statement is confirmed is just to say that one accepts it as part of one's best current theory. Hence if we adopt Quine's minimalist explication of confirmation, and give up hope of drawing the analytic–synthetic boundary in terms of a supposedly substantive, theory-independent relation of confirmation by experience, as Quine recommends, then "*S* confirmed come what may" simply amounts to "*S* is a statement of our current theory and *S* is guaranteed to be part of every theory that we will later come to accept."

Quine thinks it is worth considering, if only briefly, the question whether a boundary between analytic and synthetic sentences can be drawn by appealing to this minimalist explication of "*S* is confirmed come what may."[4] He points out, in effect, that it is enough to formulate this question clearly to see that the answer to it is "No." The problem is that all parties to the dispute about analyticity should accept

(P) No statement we now accept is guaranteed to be part of every scientific theory that we will later come to accept.

from which it follows that no statement is confirmed come what may in the minimalist sense in question.

On this reading, in paragraph 2 Quine does not assert, conversationally imply, or presuppose (R). Instead, he points out, in effect, that

---

[4] There is a passage in Quine's 1936 paper "Truth by Convention" in which he says that the statements of logic and mathematics are "destined" to be part of any scientific theory we later accept:

There are statements which we choose to surrender last, if at all, in the course of revamping our sciences in the face of new discoveries; and *among these there are some which we would not surrender at all, so basic are they to our whole conceptual scheme.* Among the latter are to be counted the so-called truths of logic and mathematics....*these statements are destined to be maintained independently of our observations of the world.* (Quine 1976, p. 102; my emphasis)

Is Quine claiming here that truth of logic and mathematics that we currently accept are *guaranteed* to be part of every scientific theory that we later come to accept? If so, he is thereby rejecting (P), and his argument in §6 of "Two Dogmas" represents a change in view. Another possibility, more likely, I believe, is that in the above passage Quine is expressing his *confidence* that we will not in fact rescind any of the statements we take to be logically true. This is a prediction, and likely a correct one, not a methodological principle that conflicts with (P).

IF

(a)  his arguments in §§1–4 are successful, and
(b)  as he argues in §5, no boundary between analytic and synthetic sentences can be drawn in terms of confirmation by experience, and
(c)  as he suggests in paragraph 1 of §6, to say that a statement is confirmed is just to say that one accepts it as part of one's best current theory,

THEN

(d)  To say of a statement which we accept now, and is therefore confirmed in Quine's minimalistic sense, that it is "confirmed come what may" is to say that it is guaranteed to be part of every scientific theory that we will later come to accept.

Quine then notes, in effect, that all parties to the dispute about analyticity should accept (P), from which it follows that no statement *S* that we now accept is 'analytic' in the proposed minimalistic sense. Since this is the *only* sense of 'analytic' that Quine can think of that has not already been ruled out by the arguments in §§1–5 of "Two Dogmas of Empiricism" and that is compatible with the minimalist account of confirmation that he sketches in paragraph 1 of §6, he concludes that if the view sketched in the first paragraph of §6 is right, "it becomes folly to seek a boundary between synthetic statements, which hold contingently on experience, and analytic statements, which hold come what may."

## 8.5  Why Carnap Accepts (P) but Rejects Quine's Conclusion

Quine had reason to be confident that Carnap, whose efforts to explicate the analytic–synthetic distinction are Quine's main focus in "Two Dogmas," would readily accept (P). Already in 1937, in *Logical Syntax of Language* (Carnap 1937), Carnap held that if a sentence of our theory logically contradicts an observation sentence we accept, then "some change must be made in the system," but "*There are no established rules for the kind of change which must be made*" (Carnap 1937, p. 317; my emphasis). He emphasizes that all rules, including those that (according to him) settle the logic of the language, and those that concern the laws of one's physical theory, "are laid down with the reservation that they may be altered as soon as it is expedient to do so" (Carnap 1937: 318). When paraphrased in the terms of Quine's minimalist account of confirmation, without any reliance on rules or on an analytic–synthetic distinction, Carnap's point becomes Quine's point that "there is [so] much latitude of choice as to what statements to reevaluate in the light of any single contrary experience" that we may reevaluate even "the statements of logical connections themselves."

Carnap reaffirms his commitment to this methodological principle in his 1963 "W. V. Quine on Logical Truth," where he writes (very generously, since, as he surely must have recalled, he made all these points Carnap 1937):

> Quine shows ... that a scientist, who discovers a conflict between his observations and his theory and who is therefore compelled to make a readjustment somewhere in the total system of science, has much latitude with respect to the place where a change is to be made. In this procedure, no statement is immune to revision, not even the statements of logic and of mathematics. There are only practical differences, and these are differences in degree, inasmuch as a scientist is usually less willing to abandon a previously accepted general empirical law than a single observation sentence, and still less willing to abandon a law of logic or of mathematics. With all this I am entirely in agreement. (Carnap 1963c, p. 921)

Thus Carnap interprets Quine's claim that no statement is immune to revision as (P) and accepts it. He apparently takes (P) to be an immediate *consequence* of the "field of force" picture of science that Quine sketches in the first paragraph of §6. Carnap therefore sees that Quine's reasoning in paragraph 2 draws consequences from points Quine makes in the previous paragraph, especially the point that adjustments can be made anywhere in case of a conflict with experience. So far so good.

In responding to Quine's reasoning in paragraph 2 of §6, however, Carnap does not acknowledge that in §4 of "Two Dogmas," Quine takes himself to have discredited Carnap's efforts to define 'analytic' and 'synthetic' in terms of rules of a language system.[5] Instead, to resist Quine's conclusion, Carnap rehearses his view that whether a sentence is analytic or not in a given language is settled by "rules of the language," and therefore "has nothing to do with" transitions "from a language $L_n$ to a new language $L_{n+1}$." He emphasizes that in his view,

> 'analytic in $L_n$' and 'analytic in $L_{n+1}$' are two different concepts. That a certain sentence $S$ is analytic in $L_n$ means only something about the status of $S$ within the language $L_n$: as has often been said, it means that that truth of

---

[5] In an earlier part of his reply to Quine, Carnap tries to respond to Quine's criticisms in §4; perhaps that is why Carnap believes that in responding to Quine's reasoning in paragraph 2 of §6, he can take for granted his method of clarifying "analytic" in terms of semantical rules.

*S* in $L_n$ is based on the meanings in $L_n$ of the terms occurring in *S*. (Carnap 1963c, p. 921)

What Carnap apparently does not see is that in the context of Quine's earlier arguments, the first paragraph of §6 sketches a minimalist account of *confirmation* that does not presuppose or require that a boundary between analytic and synthetic statements can be drawn. This new minimalist account of confirmation provides us with no way to draw a boundary between analytic and synthetic statements by examining a theory at given time, without including considerations about how it may be revised. This leaves us with one remaining proposal to consider and rule out – namely, the proposal that "*S* is analytic" means "*S* is confirmed come what may," where this amounts to "*S* is guaranteed to be part of every theory that we will later come to accept." Quine's point in paragraph 2 is that this proposal is immediately undermined by (P), which both he and Carnap accept.

## 8.6 Why Grice and Strawson Accept (P) but Reject Quine's Conclusion

Although Quine did not have Grice and Strawson in mind when he wrote "Two Dogmas," they would also readily accept (P). "The point of substance (or one of them) that Quine is making, by this emphasis on revisability," they write, "is that there is no absolute necessity about the adoption or use of any conceptual scheme whatever" (Grice and Strawson 1956, p. 157). It is clear from the context that Grice and Strawson intend to use the phrase "conceptual scheme" in a way that Quine uses it in "Two Dogmas," as, for example, when he writes of "the conceptual scheme of science," on page 44 (of Quine 1953b).[6] For Quine a conceptual scheme is simply a body of beliefs related to experience in the way he sketches in the first paragraph of §6. As we saw, revisions of a conceptual scheme are occasioned by reevaluations of some of its statements, including, in some cases, the retraction of some statements. Hence if there is no conceptual scheme that every reasonable inquirer must adopt, as Grice and Strawson grant, then (P) is true.

---

[6] In the next sentence Grice and Strawson offer another formulation of the same "point of substance" that presupposes that some statements are analytic, and they remark that their second formulation of the point is expressed "in terms [Quine] would reject" (Grice and Strawson 1956, p. 157). This implies that in their first formulation they intend to use the phrase "conceptual scheme" in a way that Quine can accept, hence, presumably, in the way he uses it in "Two Dogmas," as, for example, in Quine 1953b, on page 44.

Grice and Strawson assume that Quine's reasoning in paragraph 2 is supposed to be independent of his reasoning in earlier parts of "Two Dogmas." This assumption leads them to miss Quine's strategy of exposing the fruitlessness of the most plausible ways of clarifying "analytic" and "synthetic" and thereby showing that those who use these words have not given them any meaning. Grice and Strawson therefore search for an interpretation of Quine's claim that no statement is immune to revision according to which the claim by itself implies that there are no analytic statements.[7] In effect, they believe, on the most plausible interpretation, "no statement is immune to revision" amounts to (R). Taking this interpretation for granted, they object that Quine has not established (R), and that all he is clearly entitled to is (P), which by itself does not imply that there are no analytic statements. It should now be clear that this objection is irrelevant to Quine's argument in paragraph 2 of §6.

Grice and Strawson have another strategy for defending analyticity, however: they invite us to consider examples of statements sincere affirmations of which we fail to understand.[8] They invite us to compare, for instance,

(1)  My neighbor's three-year-old child understands Russell's theory of types.

with

(2)  My neighbor's three-year-old child is an adult.

Grice and Strawson say that if person X sincerely utters (1), we are likely to ask X for proof of X's unlikely claim, whereas, if person Y sincerely utters (2), "we would be inclined to say we just don't understand what Y is saying, and to suspect that he just does not know the meaning of some

---

[7] Chalmers 2011 makes this aspect of the standard interpretation explicit by setting aside, for the sake of his reconstruction and evaluation of Quine's argument in paragraph 2 of §6 of "Two Dogmas," every other argument that Quine offers in the paper. It was only when I read Chalmers's extreme version of the standard interpretation and tried to evaluate it that I began to realize how badly it and other less extreme versions of the standard interpretation, such as the one presented by Grice and Strawson, fit the text. Even as a challenge to (R), however, Chalmers's argument in Chalmers 2011 is unsuccessful, for reasons I explain in Chapter 9 of this volume.

[8] Grice and Strawson also challenge Quine's argument that synonymy and analyticity cannot be defined in terms of confirmation, and propose an account of their own, according to which "two statements are synonymous if and only if any experiences which, on certain assumptions about the truth values of other statements, confirm or disconfirm one of the pair, also, on the same assumptions, confirm or disconfirm the other to the same degree" (Grice and Strawson 1956: 156). In Quine 1960: 64–65, Quine explains why their efforts to define synonymy and analyticity in terms of confirmation fail.

of the words he is using" (Grice and Strawson 1956, p. 151). They take this sort of example to provide an informal explanation of the commonsense notion of analyticity.[9] For these types of explanation, "the distinction on which we ultimately come to rest is that between not believing something and not understanding something" (Grice and Strawson 1956, p. 151).

Grice and Strawson (and many others who follow them) assume, in effect, that if we cannot *make sense* of rejecting (asserting the negation of) a sentence *S*, such as "Bachelors are unmarried" or "If time flies then time flies," then *S* expresses an analytic truth – i.e., *S* is "true come what may" in a sense that contradicts (R). In effect, they take our failure to make sense rejecting a sentence *S* that we now accept as grounds for the following claim:

> (A) For at least one sentence *S* that we now accept there is no possible rational revision of the beliefs we now hold that (i) leads us, or other subject, *B*, rationally to judge, at a later time *t*, that *S* is false, and (ii) allows for a homophonic translation of *S*, as we use it now, by *S*, as we or *B*, use it at time *t*, after the revision.

Let us call this line of reasoning – from our failure to make sense of rejecting *S* to the conclusion that *S* is analytic, in the sense specified by (A) – the *argument from incomprehension*. This argument supports the standard reading of "no statement is immune from revision," as (R). For it suggests that to challenge (A), one would have to provide grounds for (R), which is logically incompatible with (A).

In *Word and Object*, Quine responds to Grice and Strawson's argument from incomprehension, indirectly, as follows:

> Sentences like 'No unmarried man is married', 'No bachelor is married', and '2 + 2 = 4' have a feel that everyone appreciates. Moreover the notion of "assent come what may" gives no fair hint of the intuition involved. One's reaction to denials of sentences typically felt as analytic has more in it of one's reaction to ungrasped foreign sentences. Where the sentence concerned is a law of logic ... dropping [it] disrupts a pattern on which the communicative use of a logical particle heavily depends. Much the same applies to '2 + 2 = 4', and even to 'The parts of the parts of the thing are parts of the thing'. The key words here have countless further contexts to anchor their usage, but somehow we feel that if our interlocutor will not

---

[9] Against this, Timothy Williamson argues that "someone may believe that normal human beings attain physical and psychological maturity at the age of three, explaining away all the evidence to the contrary by *ad hoc* hypotheses or conspiracy theories" (Williamson 2007, p. 85). Perhaps there are better examples, however. My points in the text do not depend on the success of Grice and Strawson's example (2).

agree with us on these platitudes there is no depending on him in most of the other contexts containing the terms in question. (Quine 1960, pp. 66–67)

This passage develops Quine's observation in his 1950 introduction to *Methods of Logic* that "because th[e] laws [of logic and mathematics] are so central, any revision of them is felt to be the adoption of a new conceptual scheme, the imposition of new meanings on old words" (Quine 1950, p. xiv). Now, in *Word and Object*, however, he compares "one's reaction to denials of sentences typically felt as analytic" with "one's reaction to ungrasped foreign sentences" and proposes to explain both by reflecting on maxims for translation. The basic idea is that translation proceeds in accord with a maxim that Quine states earlier in *Word and Object* as follows:

> The maxim of translation underlying all this is that assertions startlingly false on the face of them are likely to turn on hidden differences of language. This maxim is strong enough in all of us to swerve us even from the homophonic method that is so fundamental to the very acquisition and use of one's mother tongue. (Quine 1960, p. 59)

According to Quine, to say "assertions startlingly false on the face of them are likely to turn on hidden differences of language" is *not* to claim that there are assertions that are analytic in either the confirmational sense of (P) or the semantic sense of (A). Commenting on cases in which one's response to the retraction of a sentence is similar to one's response to an ungrasped foreign sentence, Quine writes, "[Such] intuitions are blameless in their way, but it would be a mistake to look to them for a sweeping epistemological dichotomy between analytic truths as byproducts of language and synthetic truths as reports on the world" (Quine 1960, p. 67).

In support of this evaluation, he cites "Two Dogmas."[10] Quine would surely not have cited "Two Dogmas" in this context if he thought that its arguments were vulnerable to the argument from incomprehension. Moreover, Quine rejects the argument from incomprehension for reasons that suggest he should not affirm (R). A central obstacle to affirming (R) is that we know of some statements, including basic laws of logic, our retraction of which would, as Quine says, "disrupt a pattern on which the communicative use of a logical particle heavily depends." We may therefore be unable, at present, to understand the assertions of a speaker who retracts such a law. But if we cannot understand the assertions of

---

[10] He also cites Quine 1963a. See Quine 1960, p. 67, note 7.

the speaker, then a fortiori we cannot interpret them homophonically into our current theory. Quine's reasons for rejecting the argument from incomprehension therefore also cast doubt on the standard interpretation, according to which he affirms (R).

Quine's rejection of the argument from incomprehension, and his adherence to the above maxim of translation, applies to high-level theorizing in physics, as well. He writes, for instance:

> In the case of wavicles … our coming to understand what the objects are is for the most part just our mastery of what the theory says about them. We do not learn first what to talk about and then what to say about it. (Quine 1960, p. 16)

Quine directly follows these remarks by tracing their consequences for an imagined discussion between two physicists about whether neutrinos have mass. He writes:

> Are they discussing the same objects? They agree that the physical theory which they initially share, the preneutrino theory, needs emendation in light of an experimental result now confronting them. The one physicist is urging an emendation which involves positing a new category of particles, without mass. The other is urging an alternative emendation which involves positing a new category of particles with mass. The fact that both physicists use the word 'neutrino' is not significant. To discern two phases here, the first an agreement as to what the objects are (viz. neutrinos) and the second a disagreement as to how they are (massless or massive), is absurd. (Quine 1960, p. 16)

I take these remarks to imply that on Quine's view, just as in the case of a basic logical law, to retract some part of our present theory of, say, neutrinos, would be to disrupt a pattern on which the communicative use of the term 'neutrino' now heavily depends. One might therefore find oneself unable to understand some of the utterances of a physicist who retracts a law of neutrinos. Such a physicist may continue to use the word form 'neutrino', but, in Quine's view, we should not translate his word form 'neutrino' homophonically.

Quine later pushed this point about translation even further, noting that "if the natives are not prepared to assent to a certain sentence in the rain, then equally we have reason not to translate the sentence as 'It is raining'" (Quine 1986f, p. 82).

Unlike logical laws, physical laws and utterances of "It's raining" are not supposed to be analytic. Hence Quine's account of failures of comprehension in terms of the above-stated maxim of translation applies not only to our so-called intuitions about analyticity, but, more generally, to any case in which "[a] native's unreadiness to assent to a certain sentence gives us reason not to construe the sentence as saying something whose truth should be obvious to the native at the time" (Quine 1986f, p. 82). Contrary to what Grice and Strawson argue, according to Quine, the phenomenon of incomprehension is not particularly relevant to analyticity, and can be explained by a maxim of translation, without any appeal to the notion of analyticity.[11]

## 8.7    Should Quine Affirm (R)?

One might still be inclined to reason as follows:

> Since Grice and Strawson in effect affirm (A), to oppose their view Quine would need to provide grounds for rejecting (affirming the negation of) (A). Given that (A) is logically incompatible with (R), the only way to provide such grounds is to provide grounds for affirming (R). Hence to answer the Grice-Strawson challenge, Quine most provide grounds for affirming (R). The standard interpretation is therefore correct, and Quine's reasoning is vulnerable in just the ways that Grice and Strawson pointed out.

This reasoning goes wrong in its assumption that to oppose Grice and Strawson's affirmation of (A), Quine needs to provide grounds for rejecting (A). In fact, to oppose Grice and Strawson's affirmation of (A), it is enough for Quine to have grounds for declining to affirm (A). Given that (A) is logically incompatible with (R), to oppose Grice and Strawson's affirmation of (A), it would be enough for Quine to have grounds for declining to reject (R).

My interpretation of Quine's reasoning in paragraph 2 of §6 does not speak directly to the question of what Quine's attitudes toward (A) and (R) should be. As we have seen, however, he has no good reason to affirm (R), given his views of translation. He should therefore *decline to affirm* (R). Should he go so far as to reject (i.e. affirm the negation of) (R)? Since (R) is obviously logically incompatible with (A), to *reject* (R) is thereby also to *affirm* (A). But Quine has no good reason to affirm (A),

---

[11] In his 1963 reply to Quine, Carnap writes that "analytic sentences cannot change their truth-value. But this characteristic is not restricted to analytic sentences; it holds also for certain synthetic sentences, e.g. physical postulates and their logical consequences" (Carnap 1963c, p. 921). Quine's response to the argument from incomprehension therefore highlights, in a very different way – without conceding that any statements are analytic – Carnap's view that the question whether we can coherently change our judgment about the truth value of a statement is not the key to understanding analyticity.

either. He should therefore *decline to affirm* (A). Given the logical incompatibility of (A) with (R), to *decline to affirm* (A) is to *decline to reject* (R). He should therefore *decline to reject* (R).

This combination of methodological attitudes is additional to, yet fully compatible with, Quine's reasoning in paragraph 2 of §6, as I explained it above. It also reveals a grain of truth in Putnam's later contributions to the discussion: even if we cannot *now* see how we could judge that a sentence S is false without changing its meaning to the point where we would no longer interpret it homophonically into our new theory, there is no guarantee that we will not *later* judge that S is false while still translating our previous uses of S into our (revised) theory homophonically.

Finally, it is also crucial to see that on the reading I sketched above, Quine is not (and should not be) committed to the claim that "for any view we can imagine circumstances in which we would give it up" (Harman 1967, p. 132). Quine is committed, instead, to (P), which is not a psychological claim about what we can imagine giving up, or retracting, but a methodological principle. Even if we are not now able to imagine a circumstance in which we would retract a particular statement of our current theory, we can see that no statement we now accept is guaranteed to be part of every scientific theory that we will later come to accept. To accept this is not to make a psychological claim about what we can now conceive, but to acknowledge both our fallibility and our commitment to retracting any statement we now accept if and when, in our pursuit of truth, we find it best to do so.

# PART IV

# Quine and Putnam

# CHAPTER 9

# Conditionalization and Conceptual Change
## Chalmers in Defense of a Dogma

In "Revisability and Conceptual Change in 'Two Dogmas of Empiricism'" (Chalmers 2011, henceforth "Revisability"), David Chalmers takes Quineans, including W. V. Quine himself, and Hilary Putnam, among others, to be committed to the following claim:

(1) For every sentence $S$ that a subject $A$ holds-true at a time $t_1$, there is a possible rational revision of the beliefs $A$ holds at $t_1$ that (a) leads $A$, or another subject $B$, rationally to judge, at some later time $t_2$, that $S$ is false, and (b) leaves the content of $S$ unchanged, so that the content of $S$, as $A$ or $B$ uses it at $t_2$, is the same as the content of $S$, as $A$ uses it at $t_1$.

Claim (1) is logically incompatible with the following widely accepted claim:

(2) Among the sentences that we hold-true now there is at least one sentence $S$ that is analytic, or "true in virtue of meaning," in a sense of "true in virtue of meaning" that implies that there is no possible rational revision of the beliefs we hold now that (a) leads us (or another subject) rationally to judge, at a later time $t$, that $S$ is false, and (b) leaves the content of $S$ unchanged, so that the content of $S$, as we (or another subject) use it at time $t$, after the revision, is the same as the content of $S$, as we use it now, before the revision.

For if (1) is true, there are no sentences that are "true in virtue of meaning" in the sense stated in (2).[1]

In §§V–VIII of "Revisability," Chalmers aims to address an opponent who is tempted to accept (or convinced of) claim (1) by examples

---

[1] In my view, Quineans should not commit themselves to claim (1). Instead, for reasons I sketch below, their attitude toward (1) should be that we do not have good grounds for rejecting it. Given this attitude toward (1), a Quinean should conclude that since (1) is logically incompatible with (2), we do not have good grounds for affirming (2). These subtleties do not matter to my evaluation of Chalmers's argument against (1), however, and I will not focus on them here. For more on these distinctions, see Chapter 8 of this volume.

that Quineans have used to challenge analyticity, but who also accepts Bayesian conditionalization. Chalmers argues that Bayesian conditionalization is a constraint on conceptual constancy, and that this constraint, together with certain plausible assumptions about a rational subject's conditional credences and "standard Bayesian considerations about evidence and updating" (Chalmers 2011, p. 399), should lead Chalmers's opponent to reject claim (1). My goal here is to explain why I think Chalmers's argument is unsuccessful.

## 9.1    Chalmers's Bayesian Constraint on Conceptual Constancy

Chalmers assumes that

> sentences are associated with unconditional and conditional credences for subjects at times. That is, for a given subject at a given time, a sentence $S$ will be associated with an unconditional credence $cr(S)$, and a pair of sentences $S$ and $T$ will be associated with conditional credence $cr(S|T)$. (Chalmers 2011, p. 400)

Chalmers's constraint on conceptual constancy amounts to the following formulation of a Bayesian principle of conditionalization:

> (CS) If a subject is fully rational, and if the subject acquires total evidence specified by $E$ between $t_1$ and $t_2$, and if the content of sentence $S$ does not change between $t_1$ and $t_2$, then $cr_2(S) = cr_1(S|E)$. (Chalmers 2011, p. 401)[2]

Chalmers assumes that his Quinean opponent can and should accept (CS). I grant this assumption, but only subject to the clarifications and qualifications that I sketch briefly in this section and describe in more detail in §§9.2–9.3.

Note first that (CS) states only a sufficient condition for $cr_2(S) = cr_1(S|E)$. It does not by itself place constraints on what kinds of updating are permissible for a fully rational subject in cases in which the contents of her sentences change. A Quinean methodological naturalist for whom the considerations that guide a person's changes in belief "are, where rational, pragmatic" (Quine 1953b, p. 46) and consist in the application of "scientific method itself, unsupported by ulterior controls" (Quine 1960, p. 23) may therefore embrace (CS) without thereby committing herself to the

---

[2] Chalmers should have also added that the content of $E$ does not change in the interval; I will take that as given in all the formulations of conditionalization that I consider in this chapter.

dubious claim that all rational changes in a person's beliefs must be the result of an application of Bayesian conditionalization.

A Quinean might nevertheless be inclined to agree with Gilbert Harman that "doing extensive updating by conditionalization ... would be too complicated in practice" (Harman 1986, p. 27). Challenges of this sort are primarily directed at the view that conditionalization is or should be an accurate description of how rational inquirers actually update their beliefs.[3] If one focuses instead on whether a rational inquirer might find it useful to *commit* herself to updating by conditionalization, however, the pragmatic considerations, while still important, are less troubling.

I assume that a Quinean may commit herself to updating by conditionalization, as constrained by a version of (CS), while knowing full well that her commitment has complicated consequences with which she cannot always easily comply. This assumption does not tell us when and how it may be reasonable to revise one's beliefs in a way that does not violate (CS), or, relatedly, how to understand the term "content," as it occurs in (CS). I suggest answers to these questions in §9.2, after I identify a hidden premise on which Chalmers's argument depends.

Assuming that Chalmers's Quinean opponents accept a version of (CS), how should we apply it to evaluate (1)? Here we face an immediate and well-known problem of bridging the gap between a subject's credences, on the one hand, and a subject's categorical attitudes, such as holding a sentence $S$ true, believing that $S$, or judging that $S$, on the other. To see the problem, consider what has come to be known as the

> Lockean Thesis[4]: There is a minimal threshold $n$, where $0 < n < 1$, such that a rational subject categorically asserts, holds-true, believes, or judges that $S$ at $t$ if and only if her credence $cr(S)$ at $t$ is equal to or greater than $n$.

This yields results that are at odds with how we tend to think about rational inference for our categorical attitudes when we are not thinking about credences. Suppose a rational subject categorically believes that $S$ and categorically believes that $S'$. If the subject does not give up either of these categorical beliefs, it is rational for her to infer, and thereby come to believe, categorically, that $S$ & $S'$. In such a case, however, it could happen that the subject's credences $cr(S)$ and $cr(S')$ are each equal to $n$

---

[3] Prominent proponents of Bayesian probabilism take such pragmatic challenges very seriously. See, for instance, Savage 1967 and Lindley 1982. For a recent survey of attempts to grapple with pragmatic challenges like Harman's without giving up Bayesian probabilism, see Seidenfeld, Schervish, and Kadane 2012.

[4] The Lockean Thesis was so named in Foley 1992.

(the minimum threshold mentioned in the Lockean Thesis), where $0 < n < 1$, and $cr(S)$ and $cr(S')$ are independent of each other in the sense that $cr(S|S') = cr(S)$.[5] Under these circumstances, if the subject is rational, her credence $cr(S \& S') = cr(S) \times cr(S')$. Since $0 < n < 1$, however, $cr(S) \times cr(S')$ must be *below* the threshold $n$. By the Lockean Thesis, then, the subject, if rational, does *not* categorically believe that $S \& S'$, contrary to our first and seemingly obvious thought that it is rational for her to believe $S \& S'$.

To engage with Chalmers's reasoning, I shall assume that claim (1) can be paraphrased without distortion in the language of Bayesian credence, with "acceptance" corresponding to "credence equal to or greater than $n$," for some threshold $n$, and "nonacceptance" to "credence less than $n$." "Rejection of $S$" we may interpret as "acceptance of the negation of $S$." We may also equate Chalmers's phrase "$cr(S|E)$ is high" with "$cr(S|E) \geq n$" and Chalmers's phrase "$cr(S|E)$ is low" with "$cr(S|E) < n$." That way we can ignore the problem just described, as Chalmers does. (In the end, I do not think we should ignore it; I do so here only be able to focus on another problem with Chalmers's reasoning.)

With these assumptions in place, we can rewrite (1) as

> (1′) For every sentence $S$ such that $cr_1(S)$ is high (not low) for subject $A$, there is a possible rational revision of the beliefs $A$ holds at $t_1$ such that, at some time $t_2$, after the revision, (a) $cr_2(S)$ is low (not high) for $A$ or for another subject $B$, and (b) the content of $S$, as $A$ or $B$ uses it at $t_2$, is the same as the content of $S$, as $A$ uses it at $t_1$.

## 9.2   Chalmers's Argument

I reconstruct Chalmers's argument against claim (1) as an argument against (1′), as follows:

> (P1) If claim (1′) is true, then any sentence can be rationally revised without a violation of (CS).

> (P2) Any sentence can be rationally revised without a violation of (CS) only if "for all rational subjects and all sentences $S$, there exists an evidence sentence $E$ such that $cr(S|E)$ is low [not high]."[6]

---

[5] For a brief discussion of this sense of independence, see page 166 of Hacking 1990.

[6] Chalmers 2011, p. 403. The consequent of (P2) should be understood as equivalent to "for all rational subjects $A$ and for all sentences $S$, at any given time $t_i$ at which $A$ has credences regarding $S$, there exists an evidence sentence $E$ such that $A$ can entertain $E$ at $t_i$ and $cr_i(S|E)$ is low (not high) for $A$." This formulation makes explicit what I take to be implied by "$cr(S|E)$ is low," namely, that $S$ and $E$ are sentences that $A$ can entertain at the time of the credence $cr$.

(P3)  For all rational subjects, "there are some truths $S$ such that $cr(S|E)$ is high [not low] for all [evidence sentences] $E$."[7]

(C)  Claim (1′) is false.          [From (P1)–(P3) and first-order logic]

The argument is valid, and I grant (P1) and (P3) for the sake of this discussion. In my view the weakest premise is (P2).

I can think of two reasons why one might find (P2) plausible. First, one might assume that from (CS) it follows that every rational revision of a given sentence is explained by Bayesian conditionalization. This assumption is mistaken, however, for the reasons I explained above. (I say more about this below.) Second, one might take for granted that a Quinean should accept that the term "content" that occurs in (CS), and that therefore determines what would count as a violation of (CS), has the same meaning as the term "content" as it appears in (1′), the target of Chalmers's criticism. This second assumption is also mistaken, as I shall now try to show.

Quine and Putnam argue, in effect, that in many cases of radical yet fully rational belief revision, it remains correct *after* the revisions to translate sentences we used earlier *homophonically* into our current theory, and to evaluate some sentences that we previously accepted as *false*. In "Carnap and Logical Truth," for instance, Quine reasons as follows:

> Suppose a scientist introduces a new term, for a certain substance or force. He introduces it by an act either of legislative definition or of legislative postulation. Progressing, he evolves hypotheses regarding further traits of the named substance or force. Suppose now that some such central hypothesis, well attested, identifies this substance or force with one named by a complex term built up of other portions of his scientific vocabulary. We all know that this new identity will figure in the ensuing developments quite on a par with the identity which first came of the act of legislative definition, if any, or on a par with the law which first came of the act of legislative postulation. Revisions, in the course of further progress, can touch any of these affirmations equally. *Now I urge that scientists, proceeding thus, are not thereby slurring over a meaningful distinction*. Legislative acts occur again and again; on the other hand a dichotomy of the resulting truths themselves into analytic and synthetic, truths by meaning postulate and truths by force of nature, has been given no tolerably clear meaning even as a methodological ideal. (Quine 1963a, p. 405)

---

[7] Chalmers 2011, p. 403. (P3) should be understood as equivalent to "For all rational subjects $A$ there are some sentences $S$ and some times $t_i$ such that $A$ has credences regarding $S$ at $t_i$, and for all evidence sentences $E$ that $A$ can entertain at $t_i$, $cr_i(S|E)$ is high (not low) for $A$." See previous note.

Quine's schematic reasoning in this paragraph is extended and further supported by Putnam's influential examples from the history of physics. Putnam observes, for instance, that our theories of geometrical space have changed since the eighteenth century, when the principles of Euclidean geometry were so fundamental to our way of thinking about physical space that we could not then conceive of any alternatives to those principles.

> Spatial locations play an obviously fundamental role in all of our scientific knowledge and in many of the operations of daily life. The use of spatial locations requires, however, the acceptance of some systematic body of geometrical theory. To abandon Euclidean geometry before non-Euclidean geometry was invented would be to "let our concepts crumble." (Putnam 1962a, p. 665)

This may seem at first to support Chalmers's reasoning, by suggesting that when we developed alternatives to Euclidean geometry for physical space, we also thereby changed the meanings of the terms and sentences that we used to describe physical space, in a sense of "change the meanings" that implies that it would be incorrect to translate those terms and sentences homophonically into our current non-Euclidean theory of physical space. Putnam rejects this description of his examples, however, as is clear from the following passage:

> [Modern physics says that] our space has variable curvature. This means that if two light rays stay a constant distance apart for a long time and then come closer together after passing the sun, we do not say that these two light rays are following curved paths through space, but we say rather that they follow straight paths and that two straight paths may have a constant distance from each other for a long time and then later have a decreasing distance from each other.... If anyone wishes to say, "Well, those paths aren't straight in the old sense of 'straight'," then I invite him to tell me *which* paths in the space near the sun are "really straight." And I guarantee that, first, no matter which paths he chooses as the straight ones, ... [they] will look crooked, act crooked, and feel crooked. Moreover, if anyone does say that certain non-geodesics are really straight paths in the space near the sun, then his decision will have to be a quite arbitrary one; and the theory that more or less *arbitrarily* selected curves near the sun are "really straight" ... would certainly *not* be a mere decision to "keep the meaning of words unchanged." (Putnam 1962a, p. 664)

Putnam concludes that our conception of physical space has changed radically since the eighteenth century, but that it is nevertheless correct to translate the terms and sentences that we used to describe physical space homophonically into our present theory of physical space, and to conclude that many of the sentences we used to accept, such as "Physical space is Euclidean," are false.[8]

If we take "what is preserved by a correct translation" to give us *one* minimally plausible grip on "content" – we may call it *translational content* – then in cases of the sort just sketched Quineans are committed to there being radical yet fully rational changes in belief without changes in translational content.

In some cases, when the time spans are short enough, the changes are intrasubjective: they are changes in how a single subject $A$ uses a sentence $S$ (or a set of sentences) at a time $t_1$ and at a later time $t_2$. In other cases, when the changes occur within a scientific community over very long periods of time, as in Putnam's Euclidean space example, the changes in belief are intersubjective: they are best described in terms of a sentence $S$ (or a set of sentences) as a subject $A$ uses it (or them) at $t_1$ and as another subject $B$ uses it (or them) at a later time $t_2$. From here on I will assume that we are working with a version of (CS) that encompass both intra- and intersubjective changes in belief.[9]

The key point for our evaluation of Chalmers's argument is this: Chalmers's premise (P2) is acceptable only if the following additional premise is true (where $A = B$ in intrasubjective cases, and $A \neq B$ in the intersubjective cases):

(*)   IF it is correct to translate a sentence $S$, as a subject $A$ used it at $t_1$, by the same sentence $S$, as $B$ uses it at $t_2$, THEN (in a sense of "content" constrained by (CS)) the content of sentence $S$, as $A$ used it at $t_1$, is the same as the content of $S$, as $B$ uses it at $t_2$.

Chalmers does not state (*), much less defend it. Should we accept it? I think not, for the following reason.

---

[8] We can accept this description of Putnam's Euclidean space case without committing ourselves to the existence of a sharp, principled distinction between changing what one means and changing what one believes. As Gilbert Harman explains, all we need to suppose is that

> we can … consider how to translate between someone's language before and after a given change in view. If the best translation is the homophonic translation, we say there has been a change in doctrine; if some other (non-homophonic) translation is better, we say there has been a change in meaning. (Harman 1994, p. 141)

[9] One could avoid this complication and think of (CS) as applying only to intrasubjective changes in belief, if one supposes (contrary to fact) that for each Quinean counterexample to analyticity that describes belief changes over a time interval, there is someone who is alive throughout the interval and who updates his or her beliefs in the way described in the counterexample. This simplifying move, with its false supposition, is of no use in the present context, however, unless we assume we do not need to make it, hence only if we assume that (CS) can be expanded to encompass both intra- and intersubjective changes in belief.

A person who is persuaded by the Quinean counterexamples described above and wishes to update her beliefs by conditionalization, as constrained by (CS), will naturally seek an explication of the term "content" in (CS) that renders (CS) compatible with the Quinean counterexamples. Such an explication should preserve what is useful about (CS) when it is understood in terms of an inquirer's faithfulness, for the year or the moment, to a commitment to update her beliefs by conditionalization. For a Quinean the best strategy for providing such an explication is to define "content" in (CS) as a kind of "conceptual role content." Gilbert Harman characterizes "conceptual role content" in terms of

> evidence, inference, and reasoning, including the impact sensory experience has on what one believes, the way in which inference and reasoning modify one's beliefs and plans, and the way beliefs and plans are reflected in action. (Harman 1974, p. 201)[10]

For present purposes, we need to characterize conceptual roles in terms of a subject's conditional and unconditional credences.[11] Fortunately, however, we do not need to provide a complete account of the conceptual roles of a subject's sentences in terms of her credences. We need only explicate the phrase "the conceptual role content of sentence $S$ is the same for $A$ at $t_1$ and $A$ (or another subject, $B$) at $t_2$," in terms of $A$'s (or $B$'s) faithfulness at $t_2$ to a *commitment* that $A$ makes at $t_1$ to updating by conditionalization.

A commitment of this kind is what Issac Levi calls a *confirmational commitment*, and is subject to a rule of *confirmational conditionalization* (Levi 1980, chapter 4, especially pp. 79–83). According to Levi, if a rational subject is faithful at $t_2$ to a confirmational commitment he made at $t_1$ and he acquires total evidence specified by $E$ between $t_1$ and $t_2$, then $cr_2$ (his

---

[10] As Harman points out (also on p. 201 of Harman 1994), his characterization of meaning as conceptual role is akin to Quine's proposal in Quine 1960, chapters 1 and 2, that we explicate the notion of the empirical meaning of a sentence for a speaker $A$ in terms of its relations to sensory stimulation and to $A$'s other sentences, as settled by $A$'s linguistic dispositions.

[11] Putnam proposes a model of "the speaker/hearer as possessing ... a subjective probability metric ..., a deductive logic, a preference ordering..., and a rule of action (e.g. "maximize the expected utility"...)" (Putnam 1978, p. 98). He emphasizes that

> not every change in 'use' (in the sense of the system, or in the sense of utterance conditions for sentences/behaviour responses to sentences) is a change in meaning. Meaning, in my view, is a course grid laid over use. I think there are different criteria for saying that there has been a change of meaning in the case of different sorts of words. (Putnam 1978, p. 98)

If we take meaning to be what is preserved in a good translation, this passage can be read as distinguishing between conceptual role content, as specified partly in terms of a person's conditional and unconditional credences, and translational content, as understood in terms of our best translational practices.

credence function at $t_2$) and $cr_1$ (his credence function at $t_1$) should be such that $cr_2(S) = cr_1(S|E)$. But a subject who adopts a confirmational commitment at $t_1$ is not thereby committed to updating his belief at $t_2$ in accord with the rule of confirmational conditionalization. For, as Levi argues, it is not irrational to abandon a confirmational commitment in favor of a different one. In other words, we should reject what Levi calls *confirmational tenacity*, according to which a rational subject "should remain faithful to the same confirmational commitment" (Levi 1980, p. 82).

I propose that we explicate the phrase "the conceptual role content of sentence $S$ is the same for $A$ at $t_1$ and $A$ at $t_2$," as "$A$ is faithful at $t_2$ to the confirmational commitment he made at $t_1$." Similarly, I propose that we explicate "the conceptual role content of sentence $S$ is the same for $A$ at $t_1$ and $B$ at $t_2$," as "$B$ is faithful at $t_2$ to the confirmational commitment that $A$ made at $t_1$." Given these explications, we may formulate (CS) as the principle that if $B$ is faithful at $t_2$ to a confirmational commitment $A$ made at $t_1$ and $B$ acquires total evidence specified by $E$ between $t_1$ and $t_2$, then $cr_2(S) = cr_1(S|E)$.[12] (In intrasubjective cases, of course, $A = B$.)

The central interest of these explications and the associated formulation of (CS) is that if we adopt them, it could happen that both

(a) the *conceptual role content* of sentence $S$, as $A$ used it at $t_1$, is different from the *conceptual role content* of sentence $S$, as $A$ (or another subject, $B$) uses it at $t_2$,

and

(b) it is correct to translate a sentence $S$, as $A$ used it at $t_1$, by the same sentence $S$, as $A$ (or another subject, $B$) uses it at $t_2$, and hence the *translational content* of $S$ does not change from $t_1$ to $t_2$.

In short, if "content" in (CS) is an abbreviation for "conceptual role content," and we adopt the explications I proposed in the previous paragraph, then we have no reason to accept (*).

---

[12] For an alternative explication of "conceptual role" in terms of credences, see Field 1977. Field assumes that "in the normal course of events people alter their probability functions in one of the standard ways – i.e. by conditionalization or by Richard Jeffrey's generalization of conditionalization" (Field 1977, p. 394) and concludes that, on his explication of "conceptual role," the conceptual roles of a person's sentences normally remain the same across time. Unlike Levi, Field does not emphasize that we can rationally abandon and revise our confirmational commitments. I prefer to specify sufficient conditions for constancy and change in "conceptual roles" of sentences in terms of Levi-style confirmational commitments, so I can take advantage of Levi's distinction between making such commitments and remaining faithful to them for a time, on the one hand, and accepting confirmational tenacity, which classifies all changes in confirmational commitment as irrational, on the other.

Without assuming (*), however, Chalmers has no grounds for (P2). For suppose we do not assume (*). Then if a radical yet fully rational revision in our beliefs coincides with a change in our confirmational commitments, yet we continue to accept homophonic translation of sentences used before the revision, we may judge that some of those sentences are false without violating (CS). We may allow that the *conceptual role* contents of those sentences have changed, without concluding that their *translational* contents have changed, hence without concluding that we cannot judge that sentences used before the revision are false.

One might think that constancies and changes in the sort of content that matters to (CS) should not be explicated completely independently of constancies and changes in *translational* content, for otherwise it could happen that $S$ has the same conceptual role content for $A$ at $t_1$ and $t_2$, even in a case in which we would *not* translate sentence $S$, as used by $A$ at $t_1$, by $S$, as used by $A$ at $t_2$. We can accommodate this thought by taking "content" in (CS) as short for "conceptual role content plus translational content."[13] The above reasoning still goes through, however. For even if "content" in the expanded version of (CS) is short for "conceptual role content plus translational content" we have no reason to accept (*), and hence no reason to accept (P2): a change in conceptual role content as I defined it above would still then amount to a change in the sort of "content" relevant to (CS), even in cases in which it would be correct to *continue* rely on homophonic translation after the change.

The moral is that Quine's and Putnam's examples and arguments put pressure on the philosophical term "content." We can plausibly respond to that pressure and accept (CS) by distinguishing between different components of content, including something like what I have called conceptual role content and translational content. Since a person's confirmational commitments change as a result of the radical yet rational belief changes described in cases of the sort that Quine and Putnam use to challenge the analytic–synthetic distinction, we may infer that in such cases the belief changes occasion changes in conceptual role contents, as we characterized such changes above, while translation continues to be homophonic across the changes. The principle of conditionalization that (CS) formulates is therefore not violated in such cases even though one rationally changes credence in a sentence $S$ (say, from a high credence in $S$ at $t_1$ to a low credence in $S$ at $t_2$) without there being an evidential sentence $E$ representing the total evidence acquired between $t_1$ and $t_2$, such that the conditional

---

[13] Field 1977 presents a hybrid view of this sort.

credence at $t_1$ for $S$, given $E$, $cr_1(S|E)$, is low. Chalmer's premise (P2) is therefore false.

Chalmers does not discuss this possible challenge to his reasoning. He does consider, however, the objection that by assuming that the content of a sentence is a proposition, he has begged the question against Quine (Chalmers 2011, pp. 405–406). His response is that however we interpret "content" in his principle (CS), "we need to require something like conceptual constancy to avoid counterexamples to principles such as conditionalization" (Chalmers 2011, pp. 405–406). This of course does not address the above criticism, which provides us with sufficient conditions for conceptual role constancy and conceptual role change that avoid counterexamples to (CS) but leave Chalmers without any grounds for his central premise, (P2). I conclude that Chalmers's argument against (1′) is unsuccessful.

## 9.3 Objections and Replies

**Objection 1** Chalmers can allow that there are some sorts of content that render (CS) false, as long as there is another kind (for example, what I have called conceptual-role content) that renders it true. Chalmers still gets the conclusion that not every sentence can be rationally revised without a change in the sort of content that renders (CS) true.

**Reply 1** As I emphasized above, the central critical purpose of the standard Quinean challenges to analyticity is to convince us that there are, or may be, some radical yet fully rational changes in belief that do, or would, lead us to see that a sentence that is supposed by proponents of the analytic–synthetic distinction to be analytic is in fact false. Suppose that in all such cases there is an associated change in the conceptual roles of an inquirer's sentences, where constancy and change in conceptual roles are explicated, as I propose, in terms of the inquirer's confirmational commitments. This leaves untouched the Quinean point that such rational changes may bring us to see that a sentence that is supposed by proponents of the analytic–synthetic distinction to be analytic is false.

**Objection 2** Given that not every sentence can be rationally revised without a change in the sort of conceptual role content that renders (CS) true, all Chalmers needs to discredit Quinean criticisms of analyticity is the additional premise that

(1) If a sentence $S$, as used by $A$ at $t$, cannot be rationally revised without any change in the conceptual role content of $S$ for $A$ at $t$, then $S$, as used by $A$ at $t$, is analytic.

A defense of analyticity can then proceed as follows. From (1), we may infer, for instance,

> (2) If the sentence 'Physical space is Euclidean', as used by scientists in the eighteenth century, cannot be rationally revised without any change in its conceptual role content, then 'Physical space is Euclidean', as used by scientists in the eighteenth century, is analytic.

From (CS) and the facts described in Putnam's Euclidean space case, it follows that

> (3) The sentence 'Physical space is Euclidean', as used by scientists in the eighteenth century, cannot be rationally revised without any change in its conceptual role content.

And from (2) and (3) it follows that

> (4) 'Physical space is Euclidean', as used by scientists in the eighteenth century, is analytic.

This refutes Putnam's supposed counterexample to the claim that 'Physical space is Euclidean', as used by scientists in the eighteenth century, is analytic.

**Reply 2** Putnam points out that we translate 'Physical space is Euclidean', as used by the scientists in the eighteenth century, into present-day English homophonically, and that 'Physical space is Euclidean', as used by us today, is not true. From these two points it follows that

> (5) 'Physical space is Euclidean', as used by scientists in the eighteenth century, is not true.

So the reasoning in Objection 2 implies that 'Physical space is Euclidean' is both analytic and not true. However, as Quine points out, "we know enough about the supposed significance of 'analytic' to know that analytic statements are supposed to be true" (Quine 1953b, p. 34). In other words, a constraint on explicating 'analytic' is

> (C) If $S$, as used by a speaker $A$ at time $t$, is analytic, then $S$, as used by $A$ at $t$, is true.

From (C) we may infer that

> (6) If 'Physical space is Euclidean', as used by scientists in the eighteenth century, is analytic, then 'Physical space is Euclidean' as used by scientists in the eighteenth century, is true.

From (4) and (6) we may infer

(7) 'Physical space is Euclidean' as used by scientists in the eighteenth century, is true.

But this contradicts (5). In short, (1) of Objection 2 results in a contradiction. We may conclude by *reductio ad absurdum* that (1) is false and hence Objection 2 is unsuccessful.

**Objection 3** In the classic examples sketched by Quine, Putnam, and others, it is not correct to rely on homophonic translation.

**Reply 3** This sort of objection has been around since the 1950s. It was defended, in one way or another, by Rudolf Carnap, Thomas Kuhn, Norman Malcolm, and many others. All of these older versions of Objection 3 were systematically challenged or rejected by Quineans. We should assume that Chalmers's intended opponents would not agree with Objection 3, or, at least, that they would have to be persuaded of it by a new argument. Moreover, to accept Objection 3 without demanding any new argument would commit one to rejecting all the classic challenges to analyticity *without* relying on (CS), hence without any need for Chalmers's argument against (1').

**Objection 4** The above criticism implies that some changes in our credences are not explained by Bayesian conditionalization. Such changes must therefore be irrational.

**Reply 4** This objection amounts to a dogmatic assertion of what Levi calls confirmational tenacity, according to which one should always remain faithful to a single confirmational commitment. As I noted above, Levi rejects confirmational tenacity, as do many other probabilists.[14] Even if confirmational tenacity were uncontroversial among probabilists, however, it would not be a good strategy to assume it when trying to persuade Quineans that their arguments against analyticity do not work. For confirmational tenacity rules out the pragmatic possibility of changing our confirmational commitments whenever we judge it useful to do so. It is therefore incompatible with the methodological principle that lies at the heart of Quinean challenges to analyticity – the principle that the considerations that guide a person's changes in belief, "are, where rational,

---

[14] See, for instance, Shimony 1970, §III, Earman 1992, Chapter 5, especially pp. 197–198. According to William Talbott,

> many Bayesians reject the Simple Principle of Conditionalization in favor of a qualified principle, limited to situations in which one does not change one's initial conditional probabilities. There is no generally accepted account of when it is rational to maintain rigid initial conditional probabilities and when it is not. (Talbott 2015, §6.2.C)

pragmatic" (Quine 1953b, p. 46) and consist in the application of "scientific method itself, unsupported by ulterior controls" (Quine 1960, p. 23).

**Objection 5** If our beliefs are even to be candidates for knowledge, they must be grounded in diachronic principles of reasoning. The suggestion that Bayesian conditionalization is not a complete account of rational changes in credences implies that our beliefs are not so grounded, and therefore cannot constitute knowledge.[15]

**Reply 5** This takes us to the heart of the issue between Chalmers and the Quineans. Must we suppose, as rationalists typically do, that knowledge is grounded in diachronic principles of reasoning? The same examples that lead Quineans to reject analyticity also lead them to doubt this principle of first philosophy. Chalmers tries to engage such Quineans by appealing to a principle, namely, (CS), a version of which many of them can accept. It turns out, however, for reasons I explained above, that this principle is compatible with the classic Quinean challenges to analyticity. Hence Chalmers's article does not present any new considerations in favor of the rationalist's dogma that knowledge must be grounded in diachronic principles of reasoning. We are left where we began, with a seemingly irresolvable impasse between rationalists and Quineans about how to think about rational inquiry.

---

[15] In response to the suggestion that one can violate conditionalization as one pleases, so that "most beliefs can be rationally revised at any moment into disbelief" – a suggestion that I am not defending here – Chalmers writes, "Given this much, it is not easy to see how beliefs can constitute knowledge at all" (Chalmers 2011, p. 412). I suspect that Chalmers's commitment to rationalism would lead him to say roughly the same thing in response to Reply 4, according to which some rational changes in credences do not violate Bayesian conditionalization, but are not explained by it, either.

# CHAPTER 10

# Truth and Transtheoretical Terms

One central theme in Hilary Putnam's work is that there is a deep connection between truth and transtheoretical terms – terms such as 'energy' and 'gold', whose references have remained the same despite fundamental changes in our beliefs.[1] Putnam points out, for instance, that

> just as the idealist regards 'electron' as theory dependent, so does he regard the semantical notions of reference and truth as theory dependent; just as the realist regards 'electron' as *trans-theoretical*, so does he regard truth and reference as trans-theoretical. (Putnam 1973, p. 198)

Noting these connections, some philosophers embrace the thesis that reference is transtheoretical because they have strongly realist intuitions about truth. Such philosophers typically assume that their intuitions about truth are in principle independent of our actual linguistic practices. In contrast, as I see it, Putnam's criticism of what he calls idealism – in particular, his criticism of logical positivism – starts with the methodological idea that our understanding of truth and reference is rooted in our actual practices of agreeing, disagreeing, evaluating assertions, and resolving disputes. If we embrace this idea, then to understand truth we need an account of reference that makes sense of actual cases in which we take ourselves to agree or disagree.

Putnam's most persuasive argument against positivism rests on actual cases in which speakers who accept very different theories apparently use the same terms to make incompatible assertions. He points out that the positivists' proposals prevent them from accepting that the speakers in these cases genuinely disagree, and concludes that the positivists don't understand truth. He observes, for example, that we disagree with some

---

[1] This theme runs through much of Putnam's work. It is present in Putnam 1962a, 1962b, 1973, 1975a, and 1975d. More recently Putnam has developed the theme into a criticism of deflationary theories of truth. See Putnam 1983d, 1985, 1991, and 1993. In Putnam 1973, Putnam notes that he picked up the phrase 'transtheoretical term' from Shapere 1969.

of the assertions that we take Niels Bohr to have made in 1911 by using the term 'electron', including the assertion that electrons have at each moment a determinate position and momentum. The positivist theories of truth and reference that Putnam opposes imply that the reference of our term 'electron' is different from the reference of Bohr's term 'electron' in 1911, so when we assert that electrons do not at each moment have a determinate position and momentum, we are not really disagreeing with an assertion Bohr made in 1911 by using his term 'electron'.[2] Putnam rejects this conclusion, and recommends instead that we accept the identifications of agreements and disagreements between speakers that we actually make in the midst of our everyday and scientific inquiries. He points out that to accept these identifications, and, in particular, to accept that we disagree with Bohr about electrons is to accept that some of our terms, including 'electron', are transtheoretical (Putnam 1973, p. 197).[3]

I will explore this connection between truth and transtheoretical terms by examining what I regard as Putnam's central objection to W. V. Quine's deflationary theory of truth and reference: that it leads to the absurd conclusion that two speakers cannot genuinely agree or disagree with each other. I will argue that the best way to see what is wrong with Quine's theory of truth and reference from Putnam's point of view is to recast this objection as a criticism of Quine's treatment of transtheoretical terms. In Putnam 1983d, 1985, 1991, and 1993, Putnam claims that his central objection to Quine's deflationary view of truth shows that truth must be a substantive property of some kind. In his *Dewey Lectures* (Putnam 1994b), he criticizes the idea that truth is a substantive property, but still maintains that "deflationism ... cannot properly accommodate the truism that certain claims about the world are (not merely assertable or verifiable but) *true.*"[4] What Putnam's central argument against Quine's deflationism shows, however, if we reconstruct it in the way I will suggest,

---

[2] This exposition of the problem needs qualification. Suppose that Bohr asserted that "there is an $x$ such that $x$ is an electron, $x$ is a sub-atomic particle, $x$ has negative charge, and $x$ has a definite momentum and position." Then we do not need to know how to translate his term 'electron' to see that this existential claim is false, provided that we can translate enough of the other terms in the sentence to settle that Bohr's assertion was false. But the positivist theories of truth and reference that Putnam opposes imply that the references of all of the descriptive terms in our current theory of subatomic particles are different from the references of all of the descriptive terms in Bohr's theory of subatomic particles. Hence if those positivist views were correct, we would be unable to infer, from accepted statements of our current theory of subatomic particles, that any of Bohr's theoretical claims about subatomic particles were false. This note responds to a comment by Art Melnick.

[3] I present a detailed reconstruction of this kind of argument, aimed at Carnap's analytic–synthetic distinction, in Ebbs 1997, chapter 6.

[4] Putnam criticizes the idea that truth is a substantive property in Lecture III (Putnam 1994b, pp. 488–517), most pointedly on p. 500; the quotation about the truism is from p. 501.

is that an account of truth and reference is satisfactory only if it accords with the identifications of agreements and disagreements between speakers that we actually make in the course of our inquiries. I will sketch a new kind of deflationism about truth and reference that meets this condition and accommodates the truism that some claims about the world are not merely assertable or verifiable but true.

## 10.1 Quine's Deflationary View of Truth and Denotation

Quine's philosophy starts with scientific naturalism, "the recognition that it is within science itself ... that reality is to be identified and described" (Quine 1981, p. 21). He counts logic among the sciences, but argues that there are no propositions, so the laws of logic must be formulated schematically, using Tarski-style definitions of truth and reference.

Quine's objection to propositions is itself an expression of his scientific naturalism. He reasons that

> if there were propositions, they would induce a certain relation of synonymy or equivalence between sentences themselves: those sentences would be equivalent that expressed the same proposition. Now my objection is ... that the appropriate equivalence relation makes no objective sense at the level of sentences. (Quine 1986f, p. 3)

That "the appropriate equivalence relation makes no objective sense at the level of sentences" is Quine's notorious thesis of the indeterminacy of translation, according to which a speaker's language can be mapped onto itself (and any other language that it translates can be mapped onto it) in a variety of inequivalent ways, each of which preserves the net association of sentences with sensory stimulation. This thesis amounts to a clarification of the consequences of Quine's austere naturalistic view of content. Quine thinks that such a mapping preserves the net association of sentences with sensory stimulation just in case the mapping would allow for "fluency of dialogue," described behavioristically. He assumes that the objective empirical content of any given sentence is exhausted by the behavioral dispositions that link it to sensory stimulation. He argues that these dispositions do not uniquely determine translation – different "translations" would pass a behavioristic test for "fluency of dialogue" – so translation between sentences is not an equivalence relation, and so (given that if there *were* propositions, translation between sentences *would be* an equivalence relation) there are no propositions.[5]

---

[5] Quine's classic statement of his indeterminacy thesis is in Quine 1960, chapter 2. In later writings he emphasizes the "fluency of dialogue" criterion for successful communication. See, for example,

If there are no propositions, the truth predicate does not apply to propositions. Instead, according to Quine, the truth predicate is "a device of disquotation" that applies to sentences (Quine 1986f, p. 12). In its application to particular sentences, it follows the disquotational pattern

(T) '_____' is true is and only if _____.

But there is no *advantage* to saying that 'Alexander conquered Persia' is true; it is easier and more direct to say that Alexander conquered Persia. So if the only use for the truth predicate were in application to particular sentences that we can directly affirm, one by one, then we could do without it.[6]

In Quine's view we need a truth predicate to formulate the laws of logic. Since there are no propositions, he reasons, we cannot formulate the logical law of excluded middle, for example, by saying that for every proposition $p$, either $p$ or not $p$. To formulate such laws without quantifying over propositions, we need a truth predicate:

> When we want to generalize on 'Tom is mortal or Tom is not mortal', 'snow is white or snow is not white', and so on, we ascend to talk of truth and of sentences, saying 'every sentence of the form '$p$ or not $p$' is true', or 'every alternation of a sentence with its negation is true' ... We ascend only because of the oblique way in which the instances over which we are generalizing are related to one another. (Quine 1986f, p. 11)

Quine concludes that logical laws are schematic generalizations that can be formulated only by using a truth predicate.

The pattern (T) is a promising first step toward defining a disquotational truth predicate that can be used to state logical generalizations. But it is well known that applications of (T) must be restricted if we are to avoid the liar paradox. One way to avoid the paradox is to adopt Alfred Tarski's approach (in Tarski 1936) to defining truth for particular

Quine 1992, p. 43. Quine has never explained how a behavioristic test for "fluency of dialogue" could be defined and applied. But on any plausible interpretation of "fluency of dialogue," in actual practice a systematically nonhomophonic translation manual from one English speaker's idiolect into another English speaker's idiolect would *not* allow for fluency of dialogue between the two speakers. English speakers typically take each other's words at face value, and would not accept that they were genuinely "communicating" with an English speaker who insisted on using a nonhomophonic translation manual. This shows that Quine's behavioristic test for "fluency of dialogue" between speakers must be seen as a test of how speakers would interact if they did *not* take each other's words at face value. It is unclear how to assess such a counterfactual.

[6] As Quine says, "So long as we are speaking only of the truth of singly given sentences, the perfect theory of truth is what Wilfred Sellars has called the disappearance theory of truth" (Quine 1986f, p. 11).

formalized languages. The approach requires that the language to which our logical schemata apply be properly regimented. A regimented first-order fragment of English, for instance, may include such sentences as '(Alexander conquered Persia) $\vee$ ¬(Alexander conquered Persia)', '$\forall x((x$ is mortal) $\rightarrow (x$ is mortal))', and '$\exists x \forall y(x$ loves $y) \rightarrow \forall y \exists x(x$ loves $y)$'. These sentences are instances, respectively, of the logical schemata '$p \vee \neg p$', '$\forall x$ (F$x \rightarrow$ F$x$)', and '$\exists x \forall y$G$xy \rightarrow \forall y \exists x$G$xy$'. We can use these schemata to state logical laws, by saying 'every sentence of the form '$p \vee \neg p$' is true', 'every sentence of the form '$\forall x$(F$x \rightarrow$ F$x$)' is true', and 'every sentence of the form '$\exists x \forall y$G$xy \rightarrow \forall y \exists x$G$xy$' is true'.

To understand these generalizations, we need a precise characterization of what counts in our language as a sentence of one of these forms, together with a clear and consistent characterization of what it means to say of any one of these sentences that it is true. The former need is met by well-known *syntactical* criteria for admissible substitutions of regimented English sentences and predicates for schematic letters,[7] and the latter need is met by Tarski's method of defining truth for particular formalized languages.

This method depends on the idea of *satisfaction* of a predicate by a sequence of objects. When the metalanguage contains the object language, a Tarski-style account of satisfaction for those predicates can be disquotational. Suppose all the variables of the object language are numbered sequentially, and let the $i$th variable in this sequence be called var($i$). A *sequence* of objects is a function from positive integers to objects; for any such sequence $s$, let $s_i$ be the $i$th object in $s$. Then if the metalanguage contains the object language, we can say, for example, that for every sequence $s$, $s$ satisfies 'red' followed by var($i$) if and only if $s_i$ is red, $s$ satisfies 'loves' followed by var($i$) and var($j$) if and only if $s_i$ loves $s_j$, and $s$ satisfies 'between' followed by var($i$), var($j$), and var($k$) if and only if $s_i$ is between $s_j$ and $s_k$.[8] Satisfaction is closely related to

---

[7] See for instance the syntactical criteria for substitution presented in chapters 26 and 28 of Quine 1982.

[8] The satisfaction clauses for predicates are needed to give inductive specifications of satisfaction conditions for sentences containing quantifiers. Suppose our regimented language contains just negation (symbolized by '¬'), alternation (symbolized by '$\vee$'), and a universal quantifier (symbolized by '$\forall$'). (In this language there is no separate symbol for the existential quantifier; existential quantifications must be expressed in terms of negation and universal quantification. Other truth functional connectives, such as '$\rightarrow$' and '$\wedge$', can be expressed in terms of '¬' and '$\vee$' in the usual way.) Then the satisfaction clauses we need, in addition to those for the $n$ simple predicates of the language, may be formulated as follows:

denotation.[9] For instance, we can say that for every sequence $s$, 'red' followed by var($i$) denotes $s_i$ if and only if $s_i$ is red, 'loves' followed by var($i$) and var($j$) denotes the ordered pair $<s_i,s_j>$ if and only if $s_i$ loves $s_j$, and 'between' followed by var($i$), var($j$), and var($k$) denotes the ordered triple $<s_i,s_j,s_k>$ if and only if $s_i$ is between $s_j$ and $s_k$.[10]

## 10.2   A First Look at Putnam's Objection

Putnam's central objection to Quine's deflationary view of truth and denotation is that it prevents Quine from seeing that speakers can agree or disagree with each other. As Putnam sees it, Quine thinks his naturalistic account of our dispositions to assent to or dissent from sentences under various prompting stimulations says all there is to say about language. Against this, Putnam argues that

> if all there is to say about language is that it consists in the production of noises (and subvocalizations) according to a certain causal pattern; if the causal story is not to be and need not be supplemented by a normative story; if there is no substantive property of *warrant* connected with the notion of *assertion*; if truth itself is not a property (the denial that truth is a property is, in fact, the central theme of all disquotational theories); then there is no way in which the noises we utter (or the subvocalizations that occur in our bodies) are more than mere "expressions of our subjectivity." (Putnam 1983d, p. 321)

($n + 1$) For all sequences $s$ and sentences S: $s$ satisfies the negation of S if and only if $s$ does not satisfy S.

($n + 2$) For all sequences $s$ and sentences S and S': $s$ satisfies the alternation of S with S' if and only if either $s$ satisfies S or $s$ satisfies S'.

($n + 3$) For all sequences $s$, sentences S, and numbers $i$: $s$ satisfies the universal quantification of S with respect to var($i$) if and only if every sequence $s'$ that differs from $s$ in at most the $i$th place satisfies S.

Suppose that together with the satisfaction clauses for the $n$ simple predicates of the language, these clauses inductively define satisfaction for all sentences of the language. Using this inductive definition of satisfaction, we can then define truth for this language as follows: a sentence of the language is *true* if and only if it is satisfied by all sequences. (The above satisfaction clauses are modeled on Quine's formulations in Quine 1986f, Chapter 3.)

[9] By "denote" I mean "true of." See Quine 1982, p. 94. Tarski characterizes truth as satisfaction by all sequences of objects. Alternatively, truth can be seen as a special case of denotation (the denotation of a 0-place predicate) as Quine explains in chapter 6 of Quine 1995. This idea was suggested much earlier, in Carnap 1942, on p. 48. See also McGee 1991, pp. 32–33.

[10] To specify the denotation of a predicate it is not necessary to identify objects as members of sequences; I do this here only to highlight the intimate connection between denotation and satisfaction.

The noises we utter count as *assertions*, hence more than mere "expressions of our subjectivity," only if in making them we can *agree* or *disagree* with other speakers. But, Putnam argues, if Quine's naturalistic account of linguistic behavior is complete from a philosophical as well as a scientific point of view, then

> we cannot genuinely disagree with each other: if I produce a noise and you produce the noise "No, that's wrong," then we have no more disagreed with each other than if I produce a noise and you produce a groan or a grunt … if I produce a noise and you produce the same noise, then this is no more *agreement* than if a bough breaks and then another breaks in the same way. (Putnam 1983d, p. 322)

Putnam concludes that "we have to recognize that there are some kind of objective properties of rightness and wrongness associated with speaking and thinking" (Putnam 1983d, p. 322). In his view, truth must be more than a device for disquotation, since we do actually make and evaluate assertions, agree and disagree with each other.[11]

## 10.3  A Reply on Quine's Behalf

Many of the sentences that we use to make statements contain no expressions that mention (refer to) linguistic expressions. For example, when we use 'Alexander conquered Persia' to state that Alexander conquered Persia, we do not mention (refer to) linguistic expressions at all. To describe the dispositions that link a speaker's sentences to impacts at her nerve endings, however, we must mention (refer to) those sentences. Hence Quine's naturalistic descriptions of linguistic behavior always mention (refer to) the sentences whose links to sensory stimulation are being described.[12]

The use–mention distinction enables Quine to make sense of disagreement or agreement in terms of logical incompatibility. If we are only describing the dispositions that link speakers' sentences to impacts at their nerve endings, the question of whether an utterance of 'Alexander conquered Persia' is logically incompatible with an utterance of '¬(Alexander

---

[11] Putnam claims that this is essentially the same as Gottlob Frege's argument against the naturalistic view that the laws of logic are psychological laws. See Frege's foreward to Frege 1893. Quine's naturalism is more sophisticated than the type of naturalism that Frege rejected, however, and Putnam's argument against Quine is accordingly less decisive than Frege's argument against the naturalisms of his day. As I will explain in §10.3, what saves Quine's naturalism from the crude mistakes that Frege exposed is that Quine respects the use–mention distinction.

[12] We cannot directly use sentences of another speaker's idiolect, so when we describe another speaker's idiolect, we can't shift our perspective and directly use that speaker's sentence. When we describe our own idiolect, in contrast, we can and do both mention and directly use our sentences.

conquered Persia)' will not arise. But we can conjoin these sentences to construct the sentence '(Alexander conquered Persia) ∧ ¬(Alexander conquered Persia)', and then *use* this sentence to express the inconsistent statement that (Alexander conquered Persia) ∧ ¬(Alexander conquered Persia). The inconsistency of the statement that (Alexander conquered Persia) ∧ ¬(Alexander conquered Persia) comes down to the inconsistency of the sentence '(Alexander conquered Persia) ∧ ¬(Alexander conquered Persia)'. To say that the sentence is inconsistent is to say that it has the form '$p \land \neg p$' and that every sentence of this form is false.[13] Let us say that two assertions (of sentences of the same language) are *logically incompatible* if the conjunction of the sentences used to make those assertions is inconsistent. Quine could then say that two assertions of sentences that are separately consistent are in *disagreement* if they are logically incompatible, and that two such assertions are in *agreement* if the conjunction of the sentence used to make one of the assertions with the negation of the sentence used to make the other assertion is inconsistent. Finally, Quine could say that two *speakers* agree or disagree according as they make assertions that are in agreement or disagreement, respectively.[14]

Quine should also reject Putnam's claim that in Quine's view, all there is to "understanding" is stated in his naturalistic description of linguistic behavior. Quine should reply that his naturalistic account of linguistic behavior is part of his theory of objective empirical content, not a replacement for or analysis of the methods of particular disciplines, such as logic, mathematics, physics, biology, and psychology. In Quine's view, there is no better way to answer questions about methods for evaluating assertions than to immerse oneself in the details of particular sciences. Quine has explored and clarified methods for evaluating assertions in mathematical logic and set theory. His central works in these areas – *Mathematical Logic* (Quine 1940), *Methods of Logic* (Quine 1982), and *Set Theory and its Logic* (Quine 1963b) – are not part of what Quine calls naturalized epistemology. They are attempts to use the vocabulary and methods of mathematical logic and set theory to clarify well-known aspects of these disciplines, and to propose new methods for them. Such

---

[13] I assume that every sentence of a properly regimented language is *false* if and only if it is *not true*. Given the standard satisfaction clauses for negation (for example, clause ($n + 1$) of note 8), every sentence of the form '$p \land \neg p$' is not true if and only if every sentence of the form '$\neg(p \land \neg p)$' is true. A simple calculation shows that every sentence of the form '$\neg(p \land \neg p)$' is true.

[14] The last three sentences of this paragraph reformulate and clarify the corresponding three sentences from the first published version of this chapter.

immersion in the details of particular sciences, however, does not yield an epistemology that is independent of or more general than the particular sciences themselves.

We can now see how misleading it is to say that in Quine's view truth is not a property. To say that a predicate "expresses a property," for Quine, is just to say that the predicate is clear and meaningful, not that there is some property it expresses. He thinks the truth predicate is clear and meaningful. It applies to 'Alexander conquered Persia', for instance, if and only if Alexander conquered Persia. We believe that Alexander conquered Persia, but we might discover that Alexander did *not* conquer Persia. The truth predicate is not simply a way of affirming a sentence, since it may not apply to a sentence even if we think it does:

> We should and do currently accept the firmest scientific conclusions as true, but when one of these is dislodged by further research we do not say that it had been true but became false. We say that to our surprise it was not true after all. Science is seen as pursuing and discovering truth rather than as decreeing it. Such is the idiom of realism, and it is integral to the semantics of the predicate 'true.' (Quine 1995, p. 67)

Trusting in the methods of particular sciences, we distinguish in practice between truth and even our firmest scientific conclusions: if we discover a mistake, we conclude that despite our best efforts, we were wrong.

These considerations show that from Quine's perspective, Putnam's objection wrongly presupposes that we can draw the use–mention distinction only if truth is a substantive property in some sense that goes beyond the observation that the truth predicate is clear and meaningful. A related mistake is to assume that if there are methods for evaluating statements, we must be able to describe these methods from a perspective independent of the particular sciences within which they are displayed and used.

## 10.4   A Reformulation of Putnam's Objection

Although Putnam certainly seems at times to suppose that we can draw the use–mention distinction only if truth is a substantive property, there is a version of Putnam's central objection to Quine that does not rest on that supposition. The basic idea is that even if we grant Quine the use–mention distinction, his indeterminacy thesis leads him to conclude that there are no genuinely transtheoretical terms, and thereby undermines his attempts to make sense of agreement and disagreement between speakers.

To appreciate this second version of Putnam's objection, it helps to distinguish between intrasubjective and intersubjective agreement or disagreement between assertions. An intrasubjective agreement or disagreement between assertions can be captured in terms of logical inconsistency in the way sketched above. Each speaker can settle for himself whether or not two of his assertions are in *disagreement* by determining whether the conjunction of the two sentences that he used to make those assertions is inconsistent. The indeterminacy thesis does not undermine this account of disagreement between assertions made by using sentences of a single idiolect. Even though each idiolect can be translated into itself in a number of inequivalent ways, negation and the other logical constants are translated in the same way for each manual according to Quine,[15] so each mapping of a speaker's idiolect onto itself that satisfies Quine's behavioristic criterion for "fluency of dialogue" will attribute the same incompatibility relations to pairs of statements made by using sentences of that idiolect.

In some apparently intrasubjective cases, however, the indeterminacy thesis undermines the commonsense assumption that agreement or disagreement between assertions is independent of our choice of a manual of translation. The reason is that, according to Quine, *nothing settles whether a speaker's idiolect at one time is the same as or different from his idiolect at another time*. Strictly speaking, for Quine, nothing settles whether the assertions that a speaker makes at one time agree or disagree with assertions she makes at another time. Such judgments must be made *relative* to a manual of translation between the speaker's earlier and later idiolects.

In this respect, according to Quine, translation between earlier and later idiolects of a single speaker is no different from translation between two idiolects of different speakers. You may assert '¬(Alexander conquered Persia)' and I may accept 'Alexander conquered Persia' but this doesn't yet settle whether we disagree, because it doesn't settle whether your sentence '¬(Alexander conquered Persia)' should be translated into my idiolect as '¬(Alexander conquered Persia)'. Our identifications of agreements and disagreements between assertions made by different speakers, or by the same speaker at different times, depend on our choice of a manual of translation that settles how sentences are mapped to sentences, and terms are mapped to terms. Bob may sincerely utter his sentence 'electrons have

[15] Quine's behavioristic criterion for determining whether a given expression $E$ is to be translated as negation, for instance, is that $E$ "turns any short sentence to which one will assent into a sentence from which one will dissent, and vice versa" Quine 1960, p. 57.

at each moment a determinate position and momentum' and I may assert that electrons do not have at each moment a determinate position and momentum; this does not determine that we disagree, even if we are both competent English speakers, since our idiolects may be mapped onto each other in different ways compatibly with all our speech dispositions. One mapping is the homophonic one, according to which Bob's sentence 'electrons have at each moment a determinate position and momentum' is translated into my idiolect as 'electrons have at each moment a determinate position and momentum'. Relative to this translation, we disagree about whether electrons have at each moment a determinate position and momentum. Yet there are other translations that do not treat 'electron' as the same word in both our idiolects. Relative to some of these alternative translations, the assertion that Bob makes by using his sentence 'electrons have at each moment a determinate position and momentum' does not disagree with my assertion that electrons do not have at each moment a determinate position and momentum.

We may summarize these points by saying that for Quine the translation of both sentences and terms is indeterminate.[16] Since the facts do not determine that the homophonic translation is the correct one, the facts do not support our judgment that 'electron' is a transtheoretical term. What look in practice like transtheoretical terms are simply terms that we translate homophonically, but that we might just as well have translated nonhomophonically.

We can now reformulate Putnam's objection as follows: Quine's indeterminacy thesis implies that our actual identifications of agreement and disagreement are dependent on arbitrary choices between equally acceptable translations between idiolects. This undermines our confidence in our actual identifications of agreement and disagreement, and thereby threatens to sever the vital link between our understanding of truth and our actual practices of agreeing, disagreeing, evaluating assertions, and resolving disputes.

According to this reformulation of Putnam's objection, Quine's theory of the relation between idiolects calls into question the practical perspective from which we typically take each other's words at face value, and thereby threatens to undermine our trust in the practices that constitute

---

[16] The key point about sentences is that

> unless pretty normally and directly conditioned to sensory stimulation, a sentence S is meaningless except relative to its own theory; meaningless intertheoretically. (Quine 1960, p. 24)

our best grip on how we evaluate statements, and when we agree or disagree. By insisting that our practice of taking some terms at face value reflects a merely subjective preference for one kind of translation manual over another, Quine discredits our practice of treating some terms as transtheoretical.

With this reconstruction of Putnam's objection in mind, let's look again at Quine's claim that truth is independent of belief. As we have seen, Quine claims that the distinction between truth and belief is "integral to the semantics of the predicate 'true'" (Quine 1995, p. 67). But from Putnam's perspective, Quine's account of the relation between idiolects prevents him from understanding the realist semantics of the predicate 'true'. If the relationship between earlier and later uses of a term is always mediated by a choice of a manual of translation, then so is our understanding of the claim that what we accept now may be false. For Quine the claim that what we accept now may be false amounts to the claim that we may *discover* that we were wrong. To make sense of such discoveries, we must imagine a translation from earlier utterances to later ones. But if attributions of truth to earlier utterances are always made relative to a choice of a translation manual, truth is not a fully objective property of those utterances.[17] Hence, according to Putnam, Quine's thesis of the relativity of our ascriptions of truth to our own past utterances prevents him from understanding the realist semantics of the predicate 'true'.

## 10.5    Putnam's Reasons for Rejecting All Deflationary Views of Truth

Quine's deflationary view of truth is a consequence of his naturalistic theory of linguistic behavior, which leads him to reject propositions, and to recommend that we use a disquotational truth predicate to state the laws of logic. All the deflationary views that Putnam criticizes – those of Paul Horwich, Richard Rorty, and Michael Williams, among others – combine behavioristic, functional, or inferential descriptions of language use with a

---

[17] Putnam writes:

> Sentences in French are true or false only relative to a translation scheme into English (or the interpreter's 'home language'). This is Quine's startling conclusion. The idea that truth and falsity are substantive properties which sentences in any language possess independently of the point of view of the interpreter must be given up. (Putnam 1985, p. 336)

In this passage Putnam does not mention Quine's indeterminacy thesis, but it is clear that he sees that thesis as one of the problematic consequences of Quine's naturalistic view of linguistic behavior.

disquotational account of truth and denotation.[18] What these deflationary views of truth have in common, according to Putnam, is that they derive the conclusion that truth is not an objective property from incomplete or inaccurate theories of linguistic activity.[19]

Putnam rejects all deflationary views for essentially the same reasons that he rejects Quine's deflationary view. Even though few of the other deflationists embrace indeterminacy, they each adopt their own theories of linguistic activity. Putnam urges us to reject all of these deflationary accounts of linguistic activity, because he thinks they prevent us from trusting our practical identifications of agreement and disagreement between speakers. Ultimately, then, Putnam's objection to deflationary views of truth and denotation has fundamentally the same structure as his objection to positivism: deflationists and positivists both make theoretical assumptions that prevent them from trusting our practical identifications of agreements and disagreements between speakers.

## 10.6 First Sketch of an Alternative Deflationary View of Truth and Denotation

My reconstruction of Putnam's objection to deflationary views of truth hints at the possibility of a deflationary view of truth and denotation that fits with our practical identifications of agreement and disagreement. To explore this possibility, I propose that we start by agreeing with Quine that the main reason we need a truth predicate is to state the laws of logic, and for this purpose all we need is a Tarski-style definition of truth and denotation for sentences and predicates of regimented languages we can use. To make sense of our practical identifications of agreement and disagreement between speakers, I propose that we put aside Quine's behavioristic descriptions of language use, and start instead with the observation that speakers of the same natural language typically take each other's words at face value without reflecting about whether they are justified in doing so. If I say, "Electrons have at each moment a determinate position and momentum,"

---

[18] For some of Putnam's criticisms of these authors, see Putnam 1983d, 1985, 1991, and 1993.
[19] In Putnam 1985, summarizing his objections to Quine's and Rorty's deflationary views of truth, and to Kripke's exposition of Wittgenstein's "skeptical solution" to the rule-following paradox, Putnam writes:

> All three tell a story about how all there is speakers and speech-dispositions, and about how we don't need any "metaphysical" notions of truth or warranted assertibility.... I say this sort of transcendental Skinnerianism has got to stop! If all there is is talk and objects internal to talk, then the idea that some pictures are "metaphysical," or "misleading," and others are not is itself totally empty. (Putnam 1985, p. 349)

you say, "Electrons do not have at each moment a determinate position and momentum," and we both take each other to be minimally competent in the use of these words, then without thinking about it, we will take each other's words at face value – you will take me to have asserted that electrons have at each moment a determinate position and momentum, and I will take you to have asserted that electrons do not have at each moment a determinate position and momentum. As a result we will take ourselves to disagree about whether electrons have at each moment a determinate position and momentum. In this case, if I am not too stubborn, and we consult the right books or scientists, I will come to agree with you.

In my view, our practice of taking each other's words at face value is integral to our understanding of truth because it is integral to our understanding of satisfaction and denotation. To see why, recall the following disquotational patterns:

(S)   For every sequence $s$, $s$ *satisfies* '___' followed by var($i$) if and only if $s_i$ is _____.

(D)   For every sequence $s$, '___' followed by var($i$) *denotes* $s_i$ if and only if $s_i$ is _____.

In principle, at least, each speaker can apply these disquotational patterns to his own words. For instance, if I affirm the results of writing 'electron' in the blanks of (S) and (D), I assert that for every sequence $s$, $s$ *satisfies* (my predicate) 'electron' followed by var($i$) if and only if $s_i$ is (an) electron, and (my predicate) 'electron' followed by var($i$) denotes an object $s_i$ if and only if $s_i$ is (an) electron.[20]

If I affirm the result of writing my word 'electron' in the blanks of (D), I can see that when I take another English speaker's word 'electron' at face value, I in effect take for granted that his word 'electron' denotes $s_i$ just in case $s_i$ is an electron, and so his word 'electron' has the same *denotation* as my word 'electron'. This is what I call a *practical judgment of sameness of denotation*. If I take another English speaker's word 'electron' at face value while I am talking to him, I make what I call a *practical judgment of sameness of denotation **at a given time***. If I take another English speaker's word 'electron' at face value while I am reading a sentence he wrote some time ago, I make what I call a *practical judgment of sameness of denotation **across time***.

---

[20] Readers who doubt that electrons are objects may prefer to think of cases in which we apply (S) and (D) to predicates that are without question true or false of objects.

Speakers of the same natural language typically take each other's words at face value without any special inquiry into whether they are justified in doing so, but they do not always take each other's words at face value in this way. For instance, if another English speaker uses a word we do not understand and cannot use, we cannot take it at face value. And sometimes another English speaker's use of an expression is so different from our use of it that we are not even tempted to take it at face value when he uses it.

We also sometimes revise and correct our unreflective ways of taking other English speakers' words. When such revisions and corrections are combined with the disquotational pattern for specifying denotations, they amount to revisions and corrections of our unreflective assumptions about the denotations of other speakers' words. That is why I call these unreflective assumptions *judgments of sameness of denotation*.

We can describe our practical judgments of sameness of denotation without thereby committing ourselves to any theory of what makes these judgments true or false. Beginning with this simple observation, I propose that we adopt a deflationary account of denotation, according to which our understanding of denotation is rooted in our practical judgments of sameness of denotation.

## 10.7   This Alternative View Contrasted with Quine's

Recall that according to my reconstruction of Putnam's central objection to Quine's deflationary view of truth, even if we accept that Quine can distinguish between the use and mention of linguistic expressions, his indeterminacy thesis implies that talk of agreement or disagreement between two speakers makes no sense unless it is relativized to a subjective choice of how to translate between their idiolects. This undermines our confidence in our practical identifications of agreement and disagreement between speakers, and thereby conflicts with our robust practical distinction between belief and truth.

In my view our understanding of agreement or disagreement is not relative to a subjective choice of how to translate between idiolects. Together with our applications of the disquotational pattern (D), our practice of taking some of our fellow English speakers' words at face value sets the parameters for our judgments about whether we agree or disagree with them. These parameters are our practical judgments of sameness of denotation. They are revisable, but nevertheless "ultimate," in the sense that there is no criterion for agreement or disagreement (or for sameness and difference of denotation) that is independent of all of them.[21]

If we accept that our practical judgments of sameness of denotation set the ultimate parameters for our understanding of sameness of denotation, we can make sense of the possibility that a conclusion we now firmly accept is false. To make sense of this possibility it is enough to imagine circumstances in which we take ourselves to have made a discovery that undermines our previous belief without undermining our practical judgments of sameness of denotation across time. This is how we can make sense of our disagreement with earlier assertions about electrons, for example.

Like Quine, I hold that the central reason we need an account of truth and denotation is to formulate the laws of logic schematically, and that for this purpose all we need are disquotational Tarski-style definitions of a truth predicate for sentences we can use. Unlike Quine, I propose that we build our practical judgments of sameness of denotation and our practical identifications of agreement and disagreement between speakers into our disquotational understanding of truth and denotation. My alternative avoids the distortions of Quine's indeterminacy thesis without committing us to an inflationary theory of truth and denotation.

## 10.8   Why Has This Alternative Been Overlooked?

Like many others, Putnam takes for granted that a deflationary theory of truth must be motivated by a commitment to something like Quine's naturalistic account of meaning and translation, which devalues the practical perspective from which we identify agreements and disagreements between speakers.[22] Quine has shown that his naturalistic account of

---

[21] Here I use Quine's idea of an "ultimate parameter" (from 1969c) to articulate my alternative to his position. It is fundamental to Quine's indeterminacy thesis that our actual linguistic interactions cannot constitute an ultimate parameter for translation or interpretation. This is a reflection of his scientific naturalism, and his observation that no scientific reconstructions of our linguistic interactions can yield a unique translation relation. My alternative rejects the assumption that a scientific reconstruction of the sort Quine offers is needed. This does not imply, as I sometimes misleadingly suggested in the first published version of this paper, that to accept the alternative I propose is to abandon scientific naturalism. Instead, to accept the alternative I propose is to reconceive of words, by taking our intra- and intersubjective practical identifications of words as ultimate parameters for our applications of disquotational definitions of truth, satisfaction, and denotation. I develop this proposal in much greater detail in Ebbs 2009.

[22] Although many philosophers acknowledge that deflationists about truth need not be committed to Quine's naturalistic view of meaning and translation, it is widely assumed that a similar sort of scientific naturalism is the only *systematic* motivation for deflationism. Even an author as thoughtful

meaning and translation is a compelling and systematic motivation for deflationism about truth. But it is not the *only possible* systematic motivation for deflationism about truth. The alternative deflationary account of truth that I recommend begins with our practice of taking each other's words at face value and described how this practice is connected with our understanding of truth and denotation.

Putnam might reply that to make sense of our linguistic practices of agreeing, disagreeing, evaluating assertions, and resolving disputes we need more than just a description of these practices; we need a substantive account of how our statements can be right or wrong.[23] It seems to me, however, that we will feel the need for such an account only if we think that to accept a deflationary view of truth is to commit oneself to a linguistic idealism according to which "the world" is merely a collection of sentences that we affirm. To rule out linguistic idealism, we may feel tempted to say such things as "When we assert a sentence such as 'Alexander conquered Persia', our assertion is substantively right or wrong, independent of what we believe"; "Reality is more than a collection of sentences held true"; or "Not everything is text." But these remarks amount to misleading attempts to elucidate aspects of the use–mention distinction for which there can be no independent argument or explanation. The use–mention distinction does not presuppose a substantive account of how it is *possible* for our statements to be true or false independently of what we believe;

---

and careful as Marian David presents only one systematic motivation for deflationism: eliminative physicalism. See David 1994, chapter 3. (This footnote, and several of the sentences in this section of the chapter, are revised from those in the original published version of the paper, to reflect my current view that to adopt the proposals in the paper is not to reject scientific naturalism, broadly understood.)

[23] This speculation about how Putnam would object to my proposal is based on some of his criticisms of disquotational theories of truth. For instance, Putnam acknowledges that

> the claim that a statement S ... never has the same *meaning* as the statement "S is assertable" is no part of the position of Williams and Horwich,

who accept deflationary views of truth; but Putnam objects that

> the statement S ... never has any kind of *substantive rightness or wrongness* beyond being assertable or having an assertable negation, on this picture. (Putnam 1991, p. 276)

Against my proposed deflationary view, Putnam would probably argue that our *trust* in practical judgments of sameness of denotation does not establish that our statements have any kind of "substantive rightness or wrongness." Putnam apparently thinks that a version of this criticism of deflationism survives his recent rejection of the idea that truth is a substantive property. In his *Dewey Lectures*, just after he criticizes the idea that truth is a substantive property, he repeats his earlier criticism that

> deflationism ... cannot properly accommodate the truism that certain claims about the world are (not merely assertable or verifiable) but *true*. (Putnam 1994b, p. 501)

that we can use our sentences to make statements that are true or false independently of what we believe is an unargued given of any plausible description of our linguistic practices.[24]

## 10.9   Truth in Other Languages

The deflationary view of truth that I propose combines disquotational specifications of satisfaction and denotation for expressions of our own language – expressions we can understand and use – with our practice of taking other speakers' words at face value. This deflationary view of truth is designed to fit with and make sense of our practices of agreeing and dis-agreeing. The motivating insight is that agreements and disagreements are typically identified in contexts in which speakers take each other's words at face value. But this insight apparently makes sense of truth only for sentences of our own language. How can this deflationary view of the connection between truth and our actual identifications of agreement and disagreement be applied to our identifications of agreements and disagree-ments we may have with speakers of other languages, such as French or German?

On any disquotational view of truth, a person can apply the truth pred-icate only to sentences that she can use. So all disquotational views of truth face a problem similar to the one I just sketched for mine. The prob-lem is to explain how a person can apply a disquotational truth predicate to sentences of languages other than her own.

The best solution to this problem is due to Quine. His idea is that for any sentence of a foreign language, such as French, we can learn how to use it, and simply extend our own language to encompass this new sen-tence. Take 'La neige est blanche', for instance. Once we learn to use this sentence, and the words that it contains, we are in a position to accept that

(t)   'La neige est blanche' is true if and only if la neige est blanche;

(s)   For every sequence $s$, $s$ *satisfies* 'blanche' followed by var($i$) if and only if $s_i$ is blanche;

(d)   For every sequence $s$, 'blanche' followed by var($i$) *denotes* $s_i$ if and only if $s_i$ is blanche.

---

[24] On this point I agree with Quine (see Chapter 5 in this volume), even though I find his descrip-tions of our linguistic behavior impoverished in the way I explain in the text.

These specifications of truth, satisfaction, and denotation for French words are given in a language that includes the French words 'La neige est blanche'.[25]

Quine's solution does not show how truth is connected with our practical identifications of agreement and disagreement. But this crucial connection is secured by the fact that just as we take utterances of English at face value, and thereby make practical judgments of sameness of denotation for expressions of English, so we take utterances of *any* expressions we can use at face value and thereby make practical judgments of sameness of denotation for those expressions, whether or not they are parts of English.

If we know that we are capable of learning to use a given sentence, then we know that we could someday be in a position to apply a disquotational truth predicate to it. If we cannot make sense of the possibility of learning to use a given sentence, however, then we cannot imagine ever being in a position to apply a disquotational truth predicate to it, and so, on the view of truth we are now considering, we cannot make sense of applying a truth predicate to it.

It also makes sense to extend the application of one's own disquotational truth predicate to the sentences of a foreign language by means of a syntactical correlation between the sentences of the foreign language and sentences of one's own idiolect. But the plausibility of extending the application of one's own truth predicate in this way depends on the plausibility of the correlation. For instance, as I argued in §10.6, Quine's criterion for such correlations undermines our practical identifications of agreement and disagreement, even between speakers of the same natural language. We should not talk of *truth relative to a correlation* unless the correlation fits with our practical judgments of sameness of denotation.[26]

But just as we have a vital practice of identifying agreements and disagreements between speakers of the same natural language, so we have a vital practice of translating between languages. Translations yield identifications of agreements and disagreements between speakers of the languages translated. Hence we may say that a sentence *S* of a foreign language is true if and only if a sentence of our own idiolect that translates *S* is disquotationally true. In practice we take established translations between sentences of two languages for granted in the same unreflective

---

[25] Quine proposed this way of applying disquotational truth to expressions of a foreign language in Quine 1953c, pp. 135–136; Field 1994 endorses this proposal.

[26] Field 1994 apparently endorses the idea of truth relative to a correlation even when the correlation does not fit with our practical judgments of sameness of denotation. I do not think such correlations license applications of 'true' to sentences of foreign languages.

way that speakers of the same language take each other's words at face value. Even in the home case, two expressions may count as the same if there are differences between them, so long as the differences are not relevant in the context. For instance, in English the predicate 'WHITE' is the same as the predicate 'white' and '*white*' when it comes to applying disquotational patterns for specifying denotation and satisfaction. We can extend this abstract idea of sameness to handle cases in which we have translations of one language into another. Relative to the accepted translation, for instance, 'blanche' is the same predicate as 'white'. This is not to say that this translation relation is *determined* by facts about the how these expressions are used. It is simply to observe that *in practice they are taken as the same predicate*. The translation of 'blanche' by 'white' is so entrenched that our judgment that a French speaker's predicate 'blanche' denotes an object *x* if and only if *x* is white is just as unreflective as our judgment that a fellow English speaker's predicate 'white' denotes an object *x* if and only if *x* is white.[27]

## 10.10    Could 'Snow Is White' Have Been False?

Putnam has raised another important objection to a Tarski-style deflationary view of truth. The heart of the objection is that the correct application 'is true' – as that predicate is pretheoretically understood – depends crucially on how speakers use the sentence to which it is applied.[28] The connection between truth and use that Putnam has in mind is highlighted by such counterfactual claims as

(C)    'snow is white' might have been used in such a way that it meant that grass is red.

This counterfactual raises an apparent difficulty for a deflationary Tarski-style definition of truth. In the circumstances described by (C), it seems

[27] Both our unreflective judgments of sameness of predicates (such as 'white' and 'blanche') and our unreflective practice of taking words of our own natural language at face value mediate our applications of disquotational patterns (S) and (D) to expressions used by other speakers. For Rudolf Carnap, such fundamental judgments of sameness of predicates, both between languages and within the same language, are analytic – true in virtue of syntactical rules. See Carnap 1937, §§2, 24–25, 62 and Carnap 1942, §§4–5. In my view these unreflective judgments are not analytic, because they are always in principle open to revision; but they are not best viewed as synthetic (factual) claims either, since we typically do not and *could* not justify them. Instead of saying that they are analytic, I prefer to say that they are *contextually a priori*, and hence always in principle open to revision. See Chapter 11 of this volume for a discussion of contextual apriority, and Ebbs 2009, chapter 4, for an extended discussion of the contextual apriority of our practical judgments of sameness of predicates.

[28] This is a paraphrase of a sentence from Putnam 1983d, p. 318.

natural to say that 'snow is white' is true if and only if grass is red. But according to the deflationary definition, 'snow is white' is true if and only if snow is white, no matter what 'snow is white' means or how it is used.

The deflationist has a straightforward answer to this objection: From the deflationist perspective, (C) amounts to the stipulation that there is (may be) a language similar to English in which the sentence 'snow is white' is to be translated by our English sentence 'grass is red'; since 'grass is red' is true in English if and only if grass is red, we can say by extension that the new sentence 'snow is white' of the stipulated language is true if and only if grass is red. This does not conflict with the disquotational scheme for 'snow is white', provided that we keep track of the different languages and sentences.

If (C) were part of a language we could learn, we could imagine learning to use the new sentence 'snow is white', and thereby extending our application of the truth predicate to include that sentence. But the example is fictional, so we would never be in a position to apply the truth predicate to the sentence in this disquotational way. We must rely instead on the bare stipulation that the sentence 'grass is red' in *our* language (English) *translates* the sentence 'snow is white' whose use is stipulated by (C).

Now our old sentence 'snow is white' must be distinguished from the new sentence 'snow is white' that is part of the language stipulated by (C). We can keep track of these different sentences by using subscripts.[29] The words 'snow' and 'white' of the language stipulated by (C) can be rewritten as 'snow$_{(C)}$' and 'white$_{(C)}$,' respectively. Thus the apparent absurdity of saying that in the circumstances described by (C), 'snow is white' is true if and only if snow is white, is just the disquotational truism that our familiar sentence 'snow is white' is true if and only if snow is white. This does not conflict with our stipulation that the sentence used in the counterfactual language – 'snow$_{(C)}$ is white$_{(C)}$' – is true if and only if grass is red; nor does it conflict with the sentence "'snow$_{(C)}$ is white$_{(C)}$' is true if and only if snow$_{(C)}$ is white$_{(C)}$" that we *would* be in a position to affirm if we *were* able to use and understand 'snow$_{(C)}$ is white$_{(C)}$'.

These considerations show that we can use a deflationary view of truth to make sense of the situation stipulated by (C), without providing a theory of the relationship between truth and use or meaning. One still might think that our pretheoretical understandings of truth and use are essentially interconnected, so Putnam is right to criticize the deflationary view

---

[29] This way of handling ambiguity is suggested by Soames 1984; see p. 427n 26.

for not fully capturing the pretheoretical understanding of truth. But this is a problem only if we suppose that the task of an account of truth is to capture all aspects of our pretheoretical understanding of truth. I doubt that there is a single underlying concept of truth that fits with all our pretheoretical assumptions about truth. Given that we can make sense of the counterfactual situation described by (C) without any account of the relationship between truth and use, I see little remaining critical force to the observation that there may be some aspects of our pretheoretical understanding of the relationship between truth and use that the deflationary account does not capture.[30]

## 10.11 Doubts about More Substantive Accounts of Truth and Denotation

I have presented a disquotational account of truth and denotation that makes sense of our practical identifications of agreement and disagreement between speakers without committing us to a substantive account of truth or denotation. I will now sketch my reasons for rejecting two initially attractive strategies for constructing substantive accounts of truth and denotation. My purpose in sketching these reasons here is to give some hint of why my view of truth and denotation is deflationary, not to prove that it is. At best, the arguments I shall sketch can help to motivate a methodological deflationism according to which we should trust our practical judgments of sameness of denotation, and doubt that there is a substantive theory that explains or grounds these judgments.

Recall that my account of truth is disquotational in the sense that it defines truth in terms of an inductive definition of satisfaction that is rooted in applications of disquotational patterns such as (S) and (D) to words of regimented languages we can use. When combined with our practice of taking other speakers' words at face value, our applications of pattern (D) result in what I call practical judgments of sameness of denotation.

I used to think that there must be a substantive account of how our language use, described in nonsemantic terms, determines the denotations of our words. I now think that if our practical judgments of sameness of denotation are taken as our ultimate parameter for judging whether individuals' words have the same denotations, then no nonsemantic

---

[30] On this point I agree with Hartry Field. See Field 1994, pp. 277–278.

description of our use of a word determines what it denotes. I became convinced of this after I constructed a number of thought experiments that challenge the assumption that the denotations of our terms are determined by the way we use them, where our 'use' of a term T is understood very broadly to include the nonsemantic relations we bear to other speakers who use T, the nonsemantic relations we bear to things in our environment to which we apply T, and the physical constitution of those things. Together with the methodological assumptions of my view, the thought experiments show that there is no criterion independent of our evolving linguistic practices for determining when two speakers agree or disagree, and which of their terms are transtheoretical.[31] If we trust our actual practical judgments of sameness of denotation more than we trust the metaphysical thesis that language use, described in nonsemantic terms, determines denotation, we will conclude that truth is not a property that sentences have or lack depending on how they are used.

Even if you accept this conclusion, however, you might think, as Putnam once did, that a sentence $S$ is true if and only if (in ideal conditions) we would be justified in asserting $S$. When we can give substantive reasons for a particular belief, we may say that it has grounds that "make" it true – that in an epistemological sense it "corresponds" with an independent "reality." Formulated for sentences, the claim is that to say that a sentence is true is to say that (in ideal conditions) we can justify the belief that we use that sentence to express. For example, if I claim that there are at least three typographical errors in a particular manuscript, you challenge me to justify this claim, and I point out three typographical errors in the manuscript, one might say that these three errors "make" the sentence 'There are at least three typographical errors in the manuscript' true.[32] Should we conclude that to say that a sentence is true is to say that (in ideal conditions) we can justify the belief that we use that sentence to express? If this generalization were true, it would amount to a substantive epistemological account of truth.

The three typographical errors "make" the sentence 'There are at least three typographical errors in the manuscript' true in the situation just described only *relative* to a background of beliefs and judgments that we take for granted in that situation. As Wittgenstein emphasizes in *On Certainty*, an epistemological sense of "correspondence" does not apply

---

[31] I argue this point in detail in Ebbs 2000, and Ebbs 2009, chapters 6 and 7.
[32] This example is based on one used in Moore 1939.

to sentences we use to express judgments that are so fundamental to our inquiries that we cannot now make sense of doubting or justifying them.[33] For instance, if we are to agree on what counts as a typographical error, we must take many unreflective judgments for granted. Among these are our practical judgments as to when two ink marks are tokens of the same letter type, which ink marks count as words, and our practical judgments of sameness of denotation. Even if unreflective judgments such as these are challenged, and we find we are unable to justify them, we may still take for granted that they are true. In this sense, truth is not a property that sentences have if and only if (in ideal conditions) we can justify the beliefs that we use them to express.

If we revise some previously entrenched belief, we do not always say that the sentence we used to express the belief has changed in truth value as a result of our revision. It may then appear that even for sentences we use to express our unreflective judgments, truth is a substantive goal of inquiry that we can grasp and understand independently of any of our beliefs or methods for evaluating assertions. But to describe examples that

---

[33] For example, Wittgenstein writes:

> One says "I know" when one is ready to give compelling grounds. "I know" relates to a possibility of demonstrating the truth. Whether someone knows something can come to light, assuming that he is convinced of it. But if what he believes is of such a kind that the grounds that he can give are no surer than his assertion, then he cannot say that he knows what he believes. (Wittgenstein 1969, §243)

Wittgenstein also observes that what someone believes is of such a kind that "the grounds that he can give are no surer than his assertion," it is misleading for him to assert that his claim is true or false. Wittgenstein is thinking of G. E. Moore's notorious claim (in Moore 1939) that he knows he has hands, for example; Moore explicitly meant this knowledge claim to "correspond" to the facts. Wittgenstein writes:

> The reason why the use of the expression "true or false" has something misleading about it is that it is like saying "it tallies with the facts or it doesn't," and the very thing that is in question is what "tallying" is here. (Wittgenstein 1969, §199)

Wittgenstein shows that in some epistemological contexts claims about which we feel very confident cannot be said to "correspond" to the facts. It makes sense to say that a given assertion "corresponds" to the facts only if we can give grounds for it. It may seem that in these passages Wittgenstein equates truth with correspondence. I read him differently, however. I think he is saying that the predicate "true" often *suggests* the idea of correspondence, and misleads us into thinking that to say a sentence is true is to say it "corresponds" to the facts. My deflationary account of truth allows us to say that a sentence is true even if we cannot provide independent grounds for believing it, hence even if we cannot say that it "corresponds" (in Wittgenstein's epistemological sense) with the facts. If one were to insist that Wittgenstein is saying that the ordinary meaning of "true" is "corresponds," then one would have to face the unwelcome consequence that assertions about which we feel most confident are neither true nor false. Aside from the strongly antirealist sound of this position, it also leaves us unable to say that an assertion about which we are now very confident may turn out to have been false.

highlight the crucial distinction between truth and belief we must always take some of our beliefs and methods for evaluating assertions for granted. In principle, any particular belief can come up for review, but not all beliefs can be reviewed at once. Truth is one thing, belief is another, but our understanding of truth always presupposes some background or other of unquestioned beliefs.

## 10.12   Conclusion

As I reconstruct it, Putnam's central objection to Quine's deflationary view of truth is that it prevents Quine from properly acknowledging actual cases of agreement and disagreement. I have explained why I think that the structure of this objection to Quine is similar to the structure of Putnam's objection to positivism. Neither objection establishes that we need an inflationary theory of truth. To show why, I sketched a new deflationary view of truth and denotation that incorporates all the practical judgments of sameness of denotation that Putnam relies on in his objections to standard deflationary views of truth, and thereby also accommodates the realist semantics of the predicate "true", without committing us to an inflationary theory of truth or denotation.

The moral of Putnam's central objection to Quine's deflationary view of truth and denotation is not that "deflationism ... cannot properly accommodate the truism that certain claims about the world are (nor merely assertable or verifiable but) *true*" (Putnam 1994b, p. 501), but that to make sense of our practical identifications of agreement and disagreement, we need to incorporate our practice of taking other speakers' words at face value into our understanding of what counts as a word of our language, and of what our words denote.[34] This practice embodies our commitment to the existence of transtheoretical terms in our language. I have argued that to make sense of our practical identifications of agreement and disagreement, and of our corresponding practical commitment to the

---

[34] It seems to me that Putnam came closer to diagnosing the problem with standard deflationary theories of truth and reference when he argued that

> the formal logic of *true* and *refers* is captured by Tarskian semantics, but the concepts of truth and reference are undetermined by their formal logic. (Putnam 1978, p. 46)

I agree with Putnam that disquotational patterns for specifying truth and denotation, which may capture the *formal* role of truth and denotation within a Tarski-style truth theory for a particular regimented language, do not by themselves show us how they should be applied to other speakers' words, or to our own past uses of words. That is one of the profound lessons of Quine's disturbing indeterminacy thesis.

existence of transtheoretical terms in our language, we need to trust our practical judgments of sameness of denotation. Once we make this shift in descriptive resources, we are in a position to develop a new kind of deflationary account of truth that incorporates our practical judgments of sameness of denotation and accommodates the truism that some claims about the world are not merely assertable or verifiable but true.

# PART V

# *Putnam*

# CHAPTER II

# Putnam and the Contextually A Priori

From its *seeming* to me – or to everyone – to be so, it doesn't follow that it
*is* so. What we can ask is whether it can make sense to doubt it.
(Ludwig Wittgenstein, *On Certainty*, §2)

When is it reasonable for us to accept a statement without evidence and
hold it immune from disconfirmation? This question lies at the heart of
Hilary Putnam's philosophy.[1] He emphasizes that our beliefs and theories
sometimes prevent us from being able to specify how a statement may
actually be false, in a sense of "specify" that goes beyond just writing or
uttering '~', 'Not', or 'It is not the case that' immediately before one writes
or utters the statement. (To save words, from here on I will assume that to
specify how a statement $S$ may actually be false, one must do more than
write or utter words that ordinarily express negation – words such as '~',
'Not', or 'It is not the case that' – immediately before one writes or utters
$S$.) In the eighteenth century, for instance, scientists did not have the theoretical understanding necessary to specify how the statement that physical space is Euclidean could be false.[2] Today, however, after Lobachevsky

---

[1] I present some of my reasons for this claim (though not the claim itself) in Ebbs 1992, and in
chapters 6, 7, and 9 of Ebbs 1997. Putnam himself once wrote: "I think that appreciating the diverse
natures of logical truths, of physically necessary truths in the natural sciences, and of what I have
for the moment lumped together under the title of framework principles – that clarifying the nature
of these diverse kinds of statements is the most important work that a philosopher can do. Not
because philosophy is necessarily about language, but because we must become clear about the roles
played in our conceptual systems by these diverse kinds of truths before we can get an adequate
global view of the world, of thought, of language, or of anything" (Putnam 1962b, p. 41).

[2] Probably no scientist in the eighteenth century would have said, "Physical space is Euclidean." We
may summarize the eighteenth century scientists' views of space in this way only if we keep in mind
that in fact their acceptance of Euclidean geometry was expressed by their commitment to such
principles as that straight lines cannot form a triangle the sum of whose angles is more than 180
degrees. We now know that this principle conflicts with the view that a straight line is a path of a
light ray, and that in some regions of space–time, paths of light rays form triangles the sum of whose
angles is more than 180 degrees. See Putnam 1962b, pp. 46–50. To save words, in the rest of this
chapter I use "the statement that physical space is Euclidean" as shorthand for a family of related
principles that we would now call Euclidean.

and Riemann discovered non-Euclidean geometries, and Einstein developed his general theory of relativity, scientists believe that physical space is non-Euclidean, and they can specify in rich detail why the statement that physical space is Euclidean is false. Generalizing from this example, we may conclude that our *current* inability to specify how a statement may actually be false does not guarantee that we will never be able to do so. Nevertheless, when we cannot specify how a statement may actually be false, it has a special methodological status for us, according to Putnam – it is *contextually a priori*.[3] In these circumstances, he suggests, it is *epistemically reasonable* for us to accept the statement without evidence and hold it immune from disconfirmation.[4]

Against this, many philosophers are inclined to reason as follows. "It is epistemically reasonable for a person to accept a particular statement only if she has epistemic grounds for accepting it. But a person's inability to specify a way in which a statement may actually be false gives her no epistemic grounds for accepting it. Therefore, if the epistemic role of the statement for her is exhausted by her inability to specify a way in which the statement may actually be false, it is not epistemically reasonable for her to accept it." Those who find this reasoning compelling typically conclude that if we want to show that it is epistemically reasonable to accept some statements without evidence, we must to try to explain how it is possible for a person to have *grounds* for accepting some statements without evidence.

In my view, however, it is more illuminating to question the idea that it is epistemically unreasonable for a person to accept *any* statement – even one that she cannot make sense of doubting – unless she has epistemic grounds for accepting it. To question this idea, I will first clarify my use of some key terminology (§11.1), present a more detailed version of the skeptical reasoning sketched in the previous paragraph (§11.2), summarize my misgivings about standard responses to it (§11.3), and explain my strategy for disarming it (§11.4). I will then examine some of Putnam's remarks about the contextually a priori (§§11.5–11.9) and argue that if a person is

---

[3] Putnam writes: "There are statements in science which can only be overthrown by a new theory – sometimes by a revolutionary new theory – and not by observation alone. Such statements *have* a sort of 'apriority' prior to the invention of the new theory which challenges or replaces them: they are *contextually a priori*. (Putnam 1983a, p. 95)

[4] As far as I know, Putnam has not used the phrase "epistemically reasonable" in this way. But his remarks about the methodological significance of contextually a priori statements suggest that he could endorse this way of expressing his view.

unable to specify any way which a statement may actually be false, she cannot make sense of the skeptic's requirement that she provide grounds for accepting it (§§11.10–11.12).

## 11.1 Three Constraints

I assume that the phrase "contextually a priori" contrasts with "contextually a posteriori." These are terms of art that can be used in different ways; one must place constraints on their use before one can raise any interesting questions about how to apply them. As I see it, ideas we associate with the words "a priori," "a posteriori," and "contextually" may guide, but do not determine, the proper use of "contextually a priori" and "contextually a posteriori": these grammatically complex terms are *logically* simple. In addition, I place three preliminary constraints on my use of "contextually a priori," "contextually a posteriori," and related epistemic terms.

The *first constraint* is that the terms "contextually a priori" and "contextually a posteriori" apply to a person's *reasons* for believing that *S* or her *entitlement* to believe that *S*, where '*S*' is replaced by a particular use in a given context of a declarative sentence.[5] (I will often use "accepting that *S*" in place of "believing that *S*," and "accept that *S*" in place of "believe that *S*." I will also assume that a particular use in a given context of a declarative sentence *S* expresses a *statement*, and that '*S*' stands in for such a statement.)

The *second constraint* is that a person has a *reason* for believing that *S* only if she can say *why* she believes that *S without presupposing that S.* (Although we sometimes say that a person has a reason for believing that *S* even if all her best attempts to explain why she believes *S* presuppose that *S*,[6] I will not use "reason" in this way.)

The *third constraint* is that a person has an *entitlement* (or is *entitled*) to believe that *S* if and only if she has no reason for believing that *S* – she cannot say why she believes that *S* without presupposing *S* – but it is

---

[5] We can define further applications of these phrases by using these primary ones. For instance, we can stipulate that a person's belief that *S* is contextually a priori (or a posteriori) for her if and only if she has a contextually a priori (or a posteriori) reason for believing or an entitlement to believe that *S*, and that *S* is contextually a priori (or a posteriori) for her if and only if she believes that *S* and her belief that *S* is contextually a priori (or a posteriori) for her.

[6] For instance, some philosophers believe that by appealing to semantical rules for using our logical connectives, we can give good reasons for accepting the inference rule modus ponens, even though we must rely on modus ponens to give those reasons. This use of "reason" is suggested by what Michael Dummett says about the justification of deductive inferences in Dummett 1978.

(epistemically) reasonable, in a sense yet to be clarified, for her to believe that S.

To highlight by contrast familiar examples of contextually a priori entitlements, I will now briefly describe examples of contextually a posteriori reasons and entitlements, and contextually a priori reasons.

Suppose you and I are watching a bird perched in a nearby tree; I say, "That's a robin," and you ask, "How do you know?" I reply, "It has a red breast." I thereby offer you a *reason* why I believe that the bird is a robin.[7] This reason does not presuppose that the bird is a robin, but provides grounds for accepting that it is a robin. Suppose I *see* that the bird has a red breast, and would not otherwise believe that it does. In this context, my reason – "It has a red breast" – is contextually a posteriori.

Now suppose that you and I both see that the bird has a red breast, I also *claim* to see that the bird as a red breast, and you challenge me to say how I know that I see that the bird has a red breast. Although it is completely obvious to me that I see that the bird has a red breast, I find I am unable to say anything persuasive or informative about why I believe this. Nevertheless, relative to the ordinary standards in that context, it seems I am *entitled* to believe that I see that the bird has a red breast even if cannot give a reason for this belief. This entitlement is contextually a posteriori.

Well-constructed proofs of logical or mathematical theorems – proofs that may presuppose special axioms and rules of inference, but do not presuppose that the theorems in question are true – are examples of contextually a priori reasons for believing the theorems.

Unlike a theorem that I can prove, however, some statements are such that I cannot say why I accept them without presupposing that they are true. For instance, I cannot say why I believe that no statement is both true and false without presupposing that no statement is both true and false. Nevertheless, in ordinary contexts it seems reasonable for me to believe this. Thus, it seems I have a contextually a priori entitlement to believe that no statement is both true and false. Similarly, as Putnam has emphasized, the belief that physical space is Euclidean was so basic for scientists in the eighteenth century that they could not say why they accepted it without presupposing it. (I will discuss this claim in more detail below.) Yet it seems that relative to the scientific standards at the time, it was reasonable for them to believe this. Thus it seems that scientists in the

---

[7] This example is modeled on J. L. Austin's goldfinch example, from his paper "Other Minds" (Austin 1946); the goldfinch example is discussed on pages 77–86.

eighteenth century had a contextually a priori entitlement to believe that physical space is Euclidean.[8]

## 11.2    A Skeptical Challenge

Beliefs that we ordinarily take for granted in giving reasons for our claims – beliefs to which we seem to be *entitled* by ordinary practice – seem especially vulnerable to skeptical challenge. Consider our confidence that we have contextually a priori entitlements to accept certain statements. I am unable to give any reasons that support my belief that no statement is both true and false, for instance, but I nevertheless take it to be reasonable to accept it. Ordinarily no one would challenge me to say why it is reasonable to accept it. But suppose someone does challenge me to say why.[9] I might reply that I cannot make sense of doubting that no statement is both true and false. But on further reflection I would realize that my inability to doubt the statement is not a reason for thinking the statement is true. At best it explains why I take it to be true. Why then do I think it is reasonable to accept the statement? I feel at a loss to answer this question, and so I begin to doubt that I have any contextually a priori entitlements, despite my initial confidence that I do.[10]

This skeptical reasoning implicitly depends on the assumption that our practices of making and evaluating statements commit us to four generalizations. The first is that

(1)    Belief does not logically imply truth.

Our commitment to this generalization is reflected in our response to the skeptical question of why we think it is reasonable to accept our belief that no statement is both true and false. We realize that we cannot adequately respond to this challenge by citing our conviction that no statement is

---

[8] Recall that I use "the statement that physical space is Euclidean" as shorthand for a family of related principles that we would now call Euclidean. See note 2.

[9] Many statements we accept are so basic to our way of thinking that we can see no point in asserting them or questioning them. It is only in the context of a skeptical challenge that we would become aware that we accept them at all. Are they genuine statements before we are aware of accepting them? The answer depends on what is meant by 'statement'. I use this word in a way that covers both acknowledged and unacknowledged commitments, where the commitments themselves are understood partly in terms of the inferences a person draws from sentences she explicitly asserts.

[10] The skeptical reasoning presented in this section resembles the "Agrippan" skepticism that Michael Williams describes on pages 61–63 in Williams 2001. It applies to contextually a posteriori entitlements, too. But the skeptical challenge to contextually a posteriori entitlements must be treated differently from the skeptical challenge to contextually a priori entitlements, so I will not address it in this chapter.

both true and false, since our conviction does not show that our acceptance of the statement is reasonable. We also realize that what counts as reasonable is intersubjective, in the sense that other participants in our search for knowledge should in principle be able to agree with us about whether it is reasonable to accept a given statement. Thus we seem committed to a second generalization about our epistemic practices:

(2) Epistemic reasons and entitlements are intersubjective.

The skeptical reasoning implicitly combines these two generalizations to suggest that

(3) It is epistemically reasonable for a person to accept a statement only if she has grounds for thinking that the statement is true.

The progression from (1) and (2) to (3) seems almost inevitable. Given (1) and (2), we cannot respond to a skeptical challenge by citing our conviction that the statement in question is true. We therefore feel we must try to explain to the skeptic why it is reasonable for us to accept the statement. But it seems that any such explanation would in effect be a reason for accepting it. In other words,

(4) A person has grounds for thinking that a statement is true only if she has reasons for accepting it.

But if we have a reason for accepting a given statement, then according to the second and third constraints of §11.1, it is not a statement that we have an *entitlement* to accept. We therefore seem forced to the conclusion that we have no contextually a priori entitlements.

## 11.3   Can We Accept (1)–(3) but Reject (4)?

Many philosophers are inclined to accept generalizations (1)–(3) but reject (4). Some would argue that even if we have no reasons for accepting a given statement, we can have grounds for taking it to be contextually a priori if the psychological processes that led us to accept it reliably yield true beliefs (Rey 1998). Others would argue that we have a capacity for "rational insight" that enables us to know directly, without reasons, that a given statement that we take to be contextually a priori is likely to be true (Bonjour 1998, Katz 1998). Yet others argue that we are entitled to accept some statements without providing any reasons for accepting them, because our acceptance of them is "constitutive" of the meanings of the words we use to express them (Boghossian 2000 and 2001, Peacocke 2000).

One problem with all of these approaches is that the skeptic of §11.2 takes (4) to be a consequence of (2), the generalization that reasons and entitlements are intersubjective. Standard ways of trying to reject (4) are not designed to convince such a skeptic,[11] from whose perspective they amount to rejections of (2), on its most natural interpretation. Yet (2) is part of the reasoning that apparently supports (3), the crucial premise in the argument that leads to the skeptical problem that these theories are supposed to solve.

Another problem is that the standard rejections of (4) tend to conflate contextually a priori entitlements with a priori entitlements. Such rejections are at best vindications of traditional examples of a priori entitlements, such as our entitlements to accept basic logical inferences or "conceptual" truths, not of Putnam's paradigm example of a contextually a priori entitlement – the entitlement of scientists in the eighteenth century to believe that physical space is Euclidean. According to the implicit meanings strategy, for instance, scientists in the eighteenth century were entitled to accept the statement that physical space is Euclidean without providing any reasons for accepting it only if their acceptance of the statement was "constitutive" of the meanings of the words they used to express it. We now know that the statement that physical space is Euclidean does not follow from the implicit meanings of the words that scientists in the eighteenth century used to express it: in the sense of meaning that is relevant to truth, we did not change the meanings of these words when we discovered that physical space is non-Euclidean. Hence the implicit meanings strategy cannot help us to avoid skepticism about such contextually a priori entitlements. Some philosophers try to make a virtue of such limitations of their epistemological theories by arguing that Putnam should not have used the word "a priori" (even qualified by "contextually") to describe the eighteenth-century scientists' attitude toward the statement that physical space is Euclidean.[12] But the important question is how we are to understand the methodological status of such statements, not whether we call them a priori.

---

[11] For instance, speaking about a skeptic who would challenge his explanation of why it is warranted for us to accept modus ponens, Paul Boghossian writes: "We cannot accept the claim that we have no warrant whatsoever for the core logical principles. We cannot conceive what such a warrant could consist in … if not in some sort of inference using those very core logical principles. So, there must be genuine warrants that will not carry any sway with a skeptic" (Boghossian 2001, p. 36). By "warrant" Boghossian means what I call "grounds." This passage therefore expresses Boghossian's choice to reject (4).

[12] See Katz 1998, p. 49. For a similar objection, but without explicit reference to Putnam, see Rey 1998, pp. 28–29.

## 11.4   My Strategy

In contrast with these standard ways of reacting to skepticism about contextually a priori entitlements, I recommend that we question whether (3) applies to all statements, including those that we take ourselves to have contextually a priori entitlements to accept. I take for granted that (3) applies to many statements that we accept. But the skeptic's implicit argument for (3) is entirely general: according to the skeptic, (3) follows inevitably from (1) and (2), and, like them, applies to all statements. Perhaps (3) does *not* follow in this way from (1) and (2). It may be that (1) and (2) hold for all statements, but (3) does not. In particular, perhaps (3) does not apply to statements that we take ourselves to have contextually a priori entitlements to accept. If (3) does not apply to these statements, then the skeptical reasoning of §11.2 depends on an overgeneralization, and the standard responses to the skeptical argument are confused and irrelevant.

My strategy is guided by the idea that a person who regards a statement *S* as contextually a priori cannot specify any way in which *S* could be false, and therefore cannot make sense of applying (3) to *S*, or of the skeptic's demand that she provide grounds for accepting *S*. To develop this idea I will explore some of Putnam's remarks about the contextually a priori. These remarks suggest an instructive but ultimately unsatisfactory reason for thinking (3) does not hold for all statements. I will explain why the reason is unsatisfactory, and then propose a better way of understanding why (3) does not hold for statements that we take ourselves to have a contextually a priori entitlement to accept.

## 11.5   Conceptual Schemes and Contextually
### a Priori Entitlements

To explain why inquirers have contextually a priori entitlements to accept certain statements, Putnam once suggested that a statement can be "necessary relative to the appropriate body of knowledge":

> When we say that a statement is necessary relative to a body of knowledge, we imply that it is included in that body of knowledge and that it enjoys a special role in that body of knowledge. For example, one is not expected to give much of a reason for that kind of statement. But we do not imply that the statement is necessarily *true*, although, of course, it is thought to be true by someone whose knowledge that body of knowledge is. (Putnam 1962a, p. 240)

Strictly speaking, Putnam should not have spoken of necessity relative to a body of knowledge, since to say that a statement is necessary or that a belief is knowledge is normally to imply that it is true. Acknowledging this point, he now recommends that we speak of "*quasi*-necessity" relative to a "conceptual scheme" (Putnam 1994c, p. 251). We must therefore ask,

(a) In what sense was the belief that physical space is Euclidean *quasi*-necessary relative to the eighteenth century scientists' conceptual scheme?

and

(b) How does this show that it was reasonable for them to accept this statement without evidence?

I will address (a) in this section and (b) in §11.6.

Putnam's answer to question (a) is that scientists in the eighteenth century could not have revised their belief that physical space is Euclidean without developing a new theory of physical space. Contextually a priori statements

> can only be overthrown by a new theory – sometimes by a revolutionary new theory – and not by observation alone.... Euclidean geometry was always revisable in the sense that no justifiable canon of scientific inquiry *forbade* the construction of an alternative geometry; but it was not always 'empirical' in the sense of having an alternative that good scientists could actually conceive. (Putnam 1983a, p. 95)

To understand this passage, one must know a little about the history of scientific theorizing about the shape of physical space from the eighteenth century until Einstein's development of the general theory of relativity.

Scientists in the eighteenth century did not distinguish between applied, or physical geometry and pure, or mathematical geometry (Reichenbach 1958, chapter 1). It was only in the nineteenth century, after Lobachevsky, Riemann, and others discovered that they could consistently describe mathematical "spaces" in which Euclid's parallel postulate does not hold, that it became possible to draw this distinction. The mathematical discovery of non-Euclidean geometries might have suggested to some that physical space may be non-Euclidean. Nevertheless, around 1830, when he first published his results, Lobachevsky called his new topic "imaginary geometry" (Boyer 1968, p. 587). Even in the late nineteenth century, after Riemann had developed non-Euclidean geometries of curved surfaces, few philosophers or mathematicians took seriously the idea that physical space is non-Euclidean. They might have regarded it as in some sense an empirical question. But the sense in which the question is empirical only became clear after Einstein changed the way we think about light and gravity. Einstein's general theory of relativity both showed how to make questions about the shape of physical space empirical and convinced many physicists and philosophers that physical space is non-Euclidean (Sklar 1974, chapters II and III).

In the eighteenth century, scientists lacked many of the conceptual resources necessary to grasp this possibility. Their failure to see any alternative to their belief that physical space is Euclidean was not based on simple oversight or ignorance. Perhaps it took longer than it might have for mathematicians and physicists to come to see how physical space could be non-Euclidean. But the eighteenth century scientists' belief that physical space is Euclidean was not epistemically irresponsible. Their understanding of geometry and physical space prevented them from seeing alternatives to Euclidean geometry, and it was no simple matter for them to overcome this obstacle. A great deal of mathematical and physical theorizing was required.

Putnam thinks there are important methodological lessons to be learned from the history of our gradual realization that questions about the shape of space are empirical. In particular, he stresses that

> before the development of general relativity theory, most people, even most scientists, could not imagine any experiences that would lead them to give up, or that would make it rational to give up, Euclidean geometry as a theory of actual space; and this is what led to the illusion that Euclidean geometry was *a priori*. (Putnam 1983b, p. 99)

By describing the methodological roles of such sentences in our rational inquiries, Putnam tries to show that some statements are so basic for us at a given time that it would not be reasonable to give them up at that time, even if we have no guarantee that they are true. He tries to convince us that if a person cannot specify any way in which a statement $S$ may be false, then she has a contextually a priori entitlement to accept $S$, even if someone else, or she herself at some later time, can specify a way in which $S$ may be false.

In short, Putnam's answer to question (a) – "In what sense was the belief that physical space is Euclidean *quasi*-necessary relative to the eighteenth century scientists' conceptual scheme?" – is that scientists in the eighteenth century had not yet developed the mathematical and physical theories that would later make it possible to specify a way in which their belief that physical space is Euclidean may actually be false. The idea is that this limitation of their "conceptual scheme" *explains* why they were unable to specify any way in which their belief that physical space is Euclidean may actually be false.

## 11.6   The Conceptual Scheme Explanation

Let us now consider question (b) – "How does this show that it was reasonable for them to accept this statement without evidence?" Note first that Putnam's explanation of why the scientists were unable to specify any way in which their belief that physical space is Euclidean may actually be false does *not* show that they had any epistemic grounds for accepting it, or that it was likely to be true. If one accepts that (3) applies to the statement that physical space is Euclidean, then one will conclude that it is epistemically irresponsible to accept the statement unless one has grounds for accepting it. Since Putnam's explanation of why the eighteenth century scientists accepted the statement strongly suggests that they had no grounds for accepting it, his explanation seems relevant only to psychology, not methodology (epistemology). Yet Putnam insists that

> the difference between statements that can be overthrown by merely conceiving of suitable experiments and statements that can be overthrown only by conceiving of whole new theoretical structures – sometimes structures, like Relativity and Quantum Mechanics, that change our whole way of reasoning about nature – is of logical and methodological significance, and not just of psychological interest. (Putnam 1962a, p. 249)[13]

How can we make sense of this?

Consider the following explanation. "Suppose we regard a statement as contextually a priori in Putnam's sense. Then our present system of beliefs – our conceptual scheme – *prevents* us from specifying any ways in which that statement may actually be false. To make sense of doubting such statements we would need to develop a new way of thinking, one that goes beyond our current understanding. But if we cannot specify any alternatives to a given statement, and no one else shows us how to do so, then that we cannot see how the statement could be false, and so we

---

[13] Putnam never changed his view on this central methodological point. In 1994, for instance, he wrote, "There are at any given time some accepted statements which cannot be overthrown merely by *observations*, but can only be overthrown by thinking of a whole body of alternative theory as well.... I insisted (and still insist) is that this is a distinction of methodological significance" (Putnam 1994c, 251).

cannot make sense of applying (3) to it. For the same reason, we cannot understand the skeptic's demand that we give grounds for accepting it. Hence Putnam's description of the role of contextually a priori statements is of methodological (epistemological) interest: it dissolves the skeptical challenge of §11.2."

This is what I will call *the conceptual scheme explanation*. I will raise three problems for it (in §§11.7–11.9), and then suggest (in §§11.10–11.12) a better way of understanding why (3) does not apply to statements we treat as contextually a priori.

## 11.7   Two Preliminary Problems for the Conceptual Scheme Explanation

The first problem is that there are statements that count as contextually a priori according to the constraints in §11.1 that are not contextually a priori according to the conceptual scheme explanation. For instance, Frege had no difficulty understanding Russell's explanation of the contradiction that arises in Frege's logic. Frege's assumption that his logic was consistent was therefore not contextually a priori according to the conceptual scheme explanation. Yet prior to Russell's letter, Frege could not specify any way in which his logic was inconsistent. He could not offer any reasons to back up his assumption that his logic was consistent, either. He was exacting and precise about all his assumptions, however, and in this sense his belief was reasonable. It therefore seems that according to the constraints in §11.1, Frege had a contextually a priori entitlement to believe that his logic was consistent. Yet according to the conceptual scheme explanation, his belief was not contextually a priori, since it did not lie deep in his system of beliefs – he could immediately see that Russell's paradox undermined his belief that his logical system was consistent.[14]

---

[14] Frege's reaction to Russell's letter was more complicated than this brief characterization suggests. In the afterword to his *Grundgesetze der Arithmetik*, vol. II (Frege 1903), Frege shows how to derive Russell's contradiction within Frege's own logical system. But, as Michael Kremer pointed out to me, Frege also suggests that the derivation shows that some expressions of his logical system have not been given any *Bedeutung*. Since Frege rejected the idea that deduction can be understood purely formally, he might have thought that his "derivations" of Russell's contradiction were not genuine derivations at all. Nevertheless, Frege found these "derivations" compelling enough to give up his basic law (V). In his letter to Russell dated June 22, 1902, six days after Russell sent Frege his famous letter about the contradiction, Frege wrote,

> Your discovery of the contradiction has surprised me beyond words and, I should almost like to say, left me thunderstruck, because it has rocked the ground on which I meant to build arithmetic. It seems … that my law V [*Grundgesetze Der Arithmetik, Vol. I*, §20] is false. (Frege 1902, p. 254)

A natural reply to this objection is that the conceptual scheme explanation concerns a slightly different topic from the topic that is implicitly defined by the constraints in §11.1. Suppose this is so. Still, the conceptual scheme explanation does not show why it was reasonable for Frege to accept that his logic was consistent. It also suggests that since there is no deep explanation of why Frege did not see that his logic was inconsistent, (3) applies to Frege's belief that his logic was consistent, so, given (4), it was *not* reasonable for him to accept it without providing reasons for it.

This suggests a second, more serious objection to the conceptual scheme explanation: we have no criterion for determining whether or not our current failure to specify a way in which a particular statement may actually be false shows that the statement is contextually a priori in the proposed sense, or whether we are just overlooking something that we would immediately recognize as a way of specifying how the statement may actually be false. Let's say that a statement is *deep* for a person if and only if she would have to develop a fundamentally new way of thinking even to conceive of how that statement may actually be false. Suppose that you are unable to specify a way in which a given statement $S$ may actually be false. The difficulty for the conceptual schemes explanation is that you cannot tell whether or not $S$ is deep for you. Tomorrow you might discover that you overlooked something, just as Frege was surprised when he read Russell's letter. But if $S$ is not deep for you, then the conceptual schemes explanation gives us no grounds for claiming that (3) does not apply to it. And if (3) does apply to it, then you are vulnerable to the skeptical reasoning presented above, because you are unable to provide any grounds for accepting the statement.

## 11.8    Two Arguments by Analogy

The most serious problem with the conceptual scheme explanation is that the imagined methodological perspective from which our statements are classified as deep or not deep for us apparently licenses an argument by analogy that would enable us to make sense of applying (3) even to statements that *are* deep for us. To understand this argument by analogy,

---

In the afterword to Frege 1903, where Frege shows how to derive the contraction within his logical system, he concludes at one point that "law (V) itself … collapses" (Frege 1903, p. 257). The change in Frege's attitude toward law (V) came about very swiftly, without the development of a fundamentally new theory of logic; in this respect it was unlike the change in attitude toward Euclidean geometry that Putnam highlights in his accounts of the contextually a priori.

it helps to consider first a simpler argument by analogy that is easier to disarm.

## First Argument by Analogy

The simpler argument may be stated as follows. "Suppose we cannot presently specify a way in which a given statement $S$ may actually be false. We nevertheless know that some statements that once seemed beyond doubt in this sense are now regarded as false. Based in our experience with such statements, we feel we understand how statements that we once regarded as beyond doubt can come to seem doubtful, and even false. By analogy with such statements, it seems that we can make sense of the possibility that $S$ is false, even though we cannot now specify any way in which $S$ may be false."

According to this simple argument by analogy, we understand the skeptic's suggestion that $S$ is possibly false, so (3) applies to $S$, and it is therefore unreasonable to accept $S$ unless we have some reason to think it is true. In this way, the argument by analogy suggests that (3) applies to *all* our statements. Once again, however, if (3) applies to all our statements, we are vulnerable to the skeptical challenge presented above: unless we can provide some reason for thinking that a given statement $S$ is true, it is unreasonable for us to accept it; we therefore have no contextually a priori entitlements.

But the argument is too simple. The natural response to it is that the analogy fails. There is a crucial difference between statements that once we could not doubt but now we can doubt, and statements that we cannot now doubt. The fact that once we could not doubt but now we can doubt a particular statement only establishes that we are fallible, and that our failure to be able to doubt a particular statement in no way guarantees it is true. This does not go beyond (1), so it does not establish that (3) applies to statements that we now regard as contextually a priori. Our fallibility does not by itself give any meaning to the claim that a particular statement may actually be false. If we cannot specify a way in which it is false, then merely mentioning our fallibility will not help us to specify a way in which it may actually be false.

## Second Argument by Analogy

A more challenging argument by analogy results when we supplement the first with a description of the methodological roles of our statements.

As we saw in §11.7, Putnam says that contextually a priori statements are "quasi-necessary" relative to a "conceptual scheme." This suggests that we can describe the methodological roles of such statements, and thereby explain why investigators regard them as "quasi-necessary" relative to their "conceptual scheme."

The appeal of this kind of explanation can be explained as follows. We now see that scientists in the eighteenth century did not simply fail to consider ways in which their statement that physical space is Euclidean may actually be false; their beliefs and theories *prevented* them from doing so. But we don't have this retrospective understanding of the centrality of any of our *current* beliefs. Tomorrow we might find that we overlooked something that we could easily have seen today. This suggests that to take ourselves have a contextually a priori entitlement to accept a given statement *S*, we must assume that *S* is deep for us, so our failure to specify ways in which *S* may actually be false is not due to a simple oversight on our part. To classify statement *S* as contextually a priori for us now is therefore to take a certain methodological perspective on our own current beliefs – to assert that our acceptance of *S* is deeply imbedded in our "system" of beliefs, and that is *why* we find ourselves unable to specify a way in which as may actually be false.

Supplemented with this methodological perspective, the first argument by analogy is transformed into a second argument by analogy that can seem more persuasive (though I will question it below). As before, the argument begins with the observation that we are now able to doubt some statements we were previously unable to doubt. Thus a contextually a priori statement may end up being doubtful, even false. The argument then continues as follows. "Statements that we currently regard as contextually a priori are from a methodological point of view no better off than those that we regarded as contextually a priori in the past. By *methodological analogy* with cases in which statements actually became doubtful, we can make sense of the possibility that a statement we now treat as contextually a priori is false, even though we cannot now specify any way in which it could be false."

Like the first argument by analogy, this argument suggests that we understand the skeptic's claim that a statement we now treat as contextually a priori may actually be false, and that it is therefore unreasonable to accept the statement unless we have some reason to think it is true. In this way, the second argument by analogy suggests that (3) applies to statements that we take to be a contextually a priori, and so the strategy of answering the skeptic by denying that (3) applies to statements we regard as contextually a priori fails.

## 11.9   Limits of the Second Argument by Analogy

Sometimes Putnam characterizes contextually a priori statements in a way
that leaves our acceptance of them vulnerable to this second argument
by analogy. In "There Is at Least One a Priori Truth" (Putnam 1983b), for
instance, Putnam contrasts contextually a priori statements with "absolutely" a priori statements – statements that "it could never be rational to
revise" – and thereby suggests that to call a statement contextually a priori
is to say that it could someday be rational to revise it (Putnam 1983b,
p. 99). This conception of contextually a priori statements suggests that
we understand how some contextually a priori statements may actually be
false, even if we cannot now *specify* a way in which they are false.[15]

Even if it is successful in some cases, however, there are limits on the
application of the second argument by analogy. As Putnam emphasizes
in Putnam 1983b, we can make no sense of the suggestion that it may be
reasonable some day to give up the minimal principle of contradiction,
according to which not every statement is both true and false. This limits
the argument by analogy by emphasizing that we do not have even the
*vaguest*, purely methodological idea of how we could end up accepting
that every statement is both true and false.[16] Hence it shows that (3) does
not apply to the minimal principle of contradiction.

By blocking the argument from analogy for the minimal principle of
contradiction, this reasoning suggests that we can discredit the skeptical
challenge in this case. Nevertheless, the second argument by analogy suggests that many of the statements that we now take ourselves to have a
contextually a priori entitlement to accept may actually be false. In this
way, the second argument by analogy suggests that (3) applies to these

---

[15] Putnam does not always describe contextually a priori statements in this way. In "Rethinking
Mathematical Necessity" (Putnam 1994c), he writes that "if we cannot *describe* circumstances
under which a belief would be falsified, circumstances under which we would be prepared to say
that –B had been confirmed, then we are not presently able to attach a clear *sense* to 'B can be
revised.' In such a case we cannot, I grant, say that B is 'unrevisable,' but neither can we intelligibly
say 'B can be revised'" (Putnam 1994c, pp. 253–254).

[16] We must keep in mind, however, that our current inability to doubt the minimal principle of
contradiction is not a reason for thinking that it is true. To make sense of asking for or providing
such a reason, we must be able to make sense of the "possibility" that the statement is not true. If
we cannot make sense of the "possibility" that every statement is both true and false, we cannot
make sense of raising any substantive question about whether the minimal principle of contradiction is true, and therefore it is a confusion to suggest that we have some reason to think it is true.
For this reason, it is misleading to say that the minimal principle of contradiction is a priori. It is
better to emphasize that we can make no sense of the "possibility" that every statement is both true
and false.

statements. This leaves us vulnerable again to the skeptical reasoning of §11.2. To accept this reasoning is to think that almost all cases in which we take ourselves to have contextually a priori entitlements are of psychological interest only, and tell us nothing about which statements it is epistemically reasonable to accept. In this way, the methodological perspective that lies behind the conceptual scheme explanation and the second argument by analogy apparently undermines the assumption that contextually a priori statements are of methodological and not only psychological interest.

## 11.10   The Second Argument by Analogy Disarmed

Despite its initial appeal, the second argument by analogy is really no better than the first one. The first one fails because the fact that we have been wrong in the past is not a reason for thinking it we are wrong now; at most it shows that we are fallible. The second analogy aims to provide an additional reason for thinking we understand how a statement that we now regard as contextually a priori may actually be false. The additional reason is suggested by the conceptual scheme explanation of why we are entitled to accept some statements as contextually a priori. According to that explanation, we are entitled to treat a statement as contextually a priori only if it is deep for us. But to make sense of the claim that a statement we now accept is deep for us, we must imagine that we can describe the "methodological role" of this statement in our current "system" of beliefs. The problem is that from our current perspective, the most we can do to clarify the methodological role of a statement that we currently accept without evidence is to search for ways of specifying how it may actually be false and report on the results of our search. Looking back on our previous beliefs, we can distance ourselves from them enough to see that in some cases we were prevented from entertaining alternatives. But we cannot take this kind of perspective on any belief that we *now* regard as contextually a priori. To imagine that we can is to imagine that we accept the belief *because* our current "conceptual scheme" prevents us from seeing any alternatives to it. But that is not a *reason* for accepting a belief. Someone *else* might be able to explain our acceptance of the belief in this way, and perhaps we will be able to explain it in that way at some future time, but right now we cannot take this perspective on it. To take this perspective on it is to undermine it. And this explains what is wrong with the second argument by analogy: the fact that we can look back on our previous beliefs and see that in some cases we were prevented from understanding alternatives to them does not show that we can make sense of the claim that our *current* beliefs prevent us from understanding alternatives to the beliefs that we now regard as beyond doubt.

The imagined methodological perspective that the second argument by analogy tries to apply to our current beliefs is an imagined third person perspective. We are tempted to think that we can take up this third-person perspective on our own current beliefs by the conceptual scheme explanation, which suggests that we cannot trust the beliefs that we now treat as contextually a priori unless we assume that they are deep for us. But the most we can coherently claim about the methodological status of beliefs we now regard as contextually a priori is that we cannot specify any ways in which they may actually be false. If we were convinced that our failure to see ways in which the statement may actually be false is due to some kind of limitation of our "conceptual scheme," we would no longer take ourselves to be entitled to accept the statement. To take ourselves to be entitled to accept the statement, however, is not to take ourselves to have some kind of guarantee that we will not find out that we are mistaken. There is a crucial distinction between admitting we are fallible, even about whether it is possible to make sense of doubting a particular statement, on the one hand, and concluding that we *understand* how the statement could actually be false, on the other. Both arguments by analogy elide this distinction.

In short, what the argument from analogy overlooks is that *to make sense of doubting a given belief one must be able to specify a particular way in which the belief may actually be false.* A corollary is that human fallibility is not by itself a reason for doubting any of our beliefs.[17] For this reason, (3) does not apply to a statement if we cannot specify any way in which it may actually be false.

At any given time we accept some statements that we cannot doubt, in the sense that we are unable to specify any ways in which they may be false. When we accept such statements, we cannot coherently distinguish between those that are revisable and those that we could never reasonably reject. Hence we cannot make sense of Putnam's suggestion (discussed in §11.9 above) that some statements are "absolutely a priori." If we cannot now specify any way in which a particular statement we accept may actually be false, we cannot be sure that we will *never* be able to make sense of giving it up without changing the topic. Nor can we be sure that we *will* someday be able to make sense of giving it up without changing the topic.

---

[17] This point is well expressed by J. L. Austin in "Other Minds." He writes that "being aware that you may be mistaken doesn't mean merely being aware that you are a fallible human being: it means that you have some concrete reason to suppose that you may be mistaken in this case" (Austin 1946, p. 98).

The most we say is that given our current understanding of the topic, *we see no way* to give up those statements without changing the topic. Since we see no way to give up those statements without changing the topic, we cannot make sense of applying (3) to them. Hence the skeptic's demand that we give grounds for these statements has no content for us.

I conclude that if a person accepts a statement *S* and she cannot specify a particular way in which *S* may actually be false, her acceptance of *S* is epistemic bedrock for her, for the moment, at least, and she cannot *make sense* of the skeptic's demand that she give grounds for accepting *S*. In these circumstances, she has what I call a contextually a priori entitlement to accept *S*.

## 11.11   Intersubjectivity

Suppose Alice can specify a way in which a given statement *S* may actually be false but Bob cannot. In these circumstances, Bob has a contextually a priori entitlement to accept *S*, but Alice does not. In effect, Alice and Bob have arrived at different conclusions about whether it is reasonable to treat *S* as contextually a priori. This seems to conflict with (2), according to which epistemic entitlements are intersubjective.

As I understand (2), however, it requires only that two inquirers should always be able to discuss and, in principle, to agree about whether a given person can specify a way in which a given statement may actually be false. When the two inquirers are discussing the beliefs of a third person, without including her in the conversation, they should be able to agree about whether she can specify a way in which a given statement may actually be false. Matters become more complicated if two inquirers together discuss the question whether one of them can specify a way in which a given statement may actually be false. Inquirers often share the background beliefs that are relevant to determining whether or not they can specify way in which a given statement may actually be false, and so they often agree about which statements to treat as contextually a priori. But there could be two inquirers, the first of whom can specify a way in which a given statement may actually be false, the second of whom cannot. The complication is that if they communicate about this, the second person may learn from the first one how to specify a way in which the statement may be false.

This can happen, for instance, in a case where one of the inquirers knows very little about a topic, and the other is a respected authority about it. Most students first learning about the shape of space are unable to doubt that it is Euclidean, because that is the only notion of physical space that they have.[18] Once they learn more geometry and physics, or read popular explanations of discoveries in these areas, they learn that it is possible to doubt that physical space is Euclidean. Before such a student learns enough to make sense of doubting that physical space is Euclidean, she has a contextually a priori entitlement to accept it. Her entitlement is intersubjective in the sense that anyone who properly understands how she thinks about physical space will see that she cannot specify any way in which that statement may actually be false.

But one might think that if she cannot specify any way in which that statement may actually be false, then she cannot *learn* from someone else that this statement is true: if she hears someone utter the words 'Physical space is not Euclidean', she should not take these words at face value; she should instead try to reinterpret them in a way that fits with what she already believes.[19]

This reasoning overlooks the fact that we typically take each other's words at face value unless we have good reason in a given context not to do so. What counts as a good reason of this kind is something that we can only discover by looking carefully at what we *actually* count as a good reason of this kind. For instance, we know that the British-English word 'football' is translated by the American-English word 'soccer'; if a British person says, "Brazil has an excellent football team," an American-English speaker has good reason not to take his term 'football' at face value. We all know, however, that a student first learning about physics typically has no good reason not to take the words of her teachers at face value, even

---

[18] One might think that a speaker could accept the statement that physical space is Euclidean, for instance, without having any idea of what physical space is or of what it is for physical space to be Euclidean, hence without being able to specify any way in which physical space may actually be Euclidean. Similarly, one might think, such a speaker can also make sense of the negation of the statement that physical space is Euclidean even though she is unable to specify a way in which the statement may actually be false. This objection overlooks the crucial point that to count as expressing the thought that physical space is Euclidean by using the sentence "Physical space is Euclidean," a speaker must be at least *minimally* competent in the use of the words that occur in the sentence, and this involves having some idea of what thought is expressed by using that sentence. I explain this point about minimal competence in Ebbs 1997, chapter 7, and Ebbs 2003. The same observations about minimal competence can also be used to correct a misunderstanding of "understanding." It might seem that to understand a statement that *p* one must be able to distinguish circumstances in which *p* is true from circumstances in which *p* is false. But this would imply that to understand the statement that *p* one must be able to specify a way in which *p* may be false, and therefore that no one understands any statements that I call contextually a priori. This overlooks our ordinary criteria for taking a person to be competent in the use of a sentence, hence to have at least a minimal understanding of the thoughts she expresses by using it.

[19] Donald Davidson's principle of charity leads inevitably to this unacceptable conclusion (despite his occasional claims to the contrary), for reasons I explain in Ebbs 2002.

if prior to her studies, she cannot make sense of doubting that space is Euclidean. Similarly, scientists who discover radically new theories typically take themselves to be talking about the same topics into which they were inquiring before they came up with their new theories. In practical terms, this means that scientists often take themselves to use the same words with the same denotations, both at a given time and over time. Speakers of the same language typically take each other's words at face value in this way. Sometimes, as in the 'football' example, there's good reason not to do so. But a person's present inability to make sense of doubting a given statement $S$ is not by itself a good reason for her to refuse to take at face value a fellow speaker's utterance of the negation of $S$.[20] She will naturally want an explanation of how $S$ could be false. Once she hears the explanation, however, she may begin to see how $S$ could be false, and she may even be convinced that $S$ *is* false.[21]

This observation should also dispel the worry that if a speaker has a contextually a priori entitlement to accept a statement $S$, then her acceptance of $S$ cannot be challenged or criticized. One might think that by *relativizing* our understanding of when a person has a contextually a priori entitlement to accept a statement $S$ to our understanding of whether she can specify a way in which $S$ may actually be false, we commit ourselves to an *epistemological relativism* according to which some statements are so basic to a person's way of looking at the world that no one who does not accept those statements can challenge or criticize her acceptance of them. As the example of the student who learns about non-Euclidean geometry and the general theory of relativity shows, however, a person may *lose* her contextually a priori entitlement to accept a statement $S$ by learning how to specify ways in which $S$ may actually be false. In this case, we see that at one time she had a contextually a priori entitlement to accept that physical space is Euclidean, because at that time she could not specify any way in which the statement could be false. Now that she is learned more, however, she no longer has a contextually a priori entitlement to accept

---

[20] The relationship between what we can make sense of doubting, and what we can understand another person to have said is subtle and context-sensitive. I know of no easy generalizations about this relationship. To understand it better, one would need to look carefully at what we would say in practice to a wide range of cases in which the limits of what a person can make sense of doubting are apparently challenged or extended by what another person writes or says.

[21] Similarly, a mathematics student who is unable to see how a certain solution to a mathematical problem she is trying to solve could possibly be correct is not thereby entitled to believe that the proposed solution is incorrect. If the proposed solution comes from a trusted and authoritative source, such as an accomplished mathematician or a respected textbook, she should suspend her belief until she understands it better.

that physical space is Euclidean. Whether or not she has a contextually a priori entitlement at a given time to accept a statement $S$ depends on whether at that time she can specify a way in which $S$ may be false. The relativity about whether or not a given speaker has a contextually a priori entitlements to accept a sentence $S$ does not insulate her acceptance of $S$ from all challenges or criticisms, since she may have a contextually a priori entitlement at one time to accept a given statement $S$, and later lose that entitlement, because she has learned how to specify a way in which $S$ may be false.

This sketch of the sense in which our practice of attributing entitlements is intersubjective partly elucidates (2), according to which reasons and entitlements are intersubjective. It also partly elucidates (1), by reminding us that we are fallible, and that our acceptance of a belief does not make it true. At the same time, however, we have concluded that if a speaker cannot specify a way in which a given statement may actually be false, then she cannot apply (3) to it, and she cannot make sense of the skeptic's demand that she provide grounds for accepting it. We can therefore accept (4) – the observation that one has grounds for accepting a given statement $S$ only if one has reasons for accepting it – without committing ourselves to skepticism about contextually a priori entitlements.

## 11.12   Conclusion

The conceptual scheme explanation creates the confused impression that we can *explain* why it is reasonable for us to accept some statements by saying that they are deep for us – that we would not be able even to make sense of giving up the statements unless we developed a new way of thinking about them. The trouble is that we cannot make sense of the claim that a sentence we *now* accept is deep for us. We might be able to make sense of this at some point in the future. But to say that a sentence is deep for us is to say that our failure to be able to specify a way in which it is false is explained by a *limitation* of our current conceptual scheme. And if we were convinced that our current acceptance of the statement is explained by a conceptual limitation of this kind, we would no longer accept it.

Our discovery of our own previous conceptual limitations shows us that our estimations of whether it is possible to specify a way in which a given statement may actually be false are fallible. But this fallibility by itself does not give any content to the claim that a given statement that we now accept is deep for us. Without any understanding of this explanatory claim, we cannot make sense of the analogy that is supposed to give content to the conceptual scheme explanation. We cannot make sense of the imagined perspective from which our current beliefs, taken together, constitute a "conceptual scheme" with built-in limitations on our ability to specify ways in which some of our statements may actually be false. For the same reason, we cannot make sense of the idea (discussed in §11.9 above) that among the statements we now accept there are some that we could *never* reasonably reject. This idea has content only if we can contrast such statements with other statements that are deep for us, but revisable. Since we cannot make sense of the idea that some statements we now accept are deep for us, we cannot make sense of the idea that some statements we now accept are "absolutely a priori," in the sense that we could never reasonably reject them.

The moral is that we cannot *explain* or *justify* our current contextually a priori entitlements. To have such entitlements is just to rely on statements that we find ourselves unable to doubt. In many cases this reliance is unreflective, yet reasonable: if challenged, we would not be able to make any sense of the possibility that the statement is false. In other cases, we persistently search for ways of specifying how a statement that we accept may be false, and fail to find any. Since we fail to find any, we cannot make sense of applying (3) to the statement, and we cannot make sense of the skeptic's demand that we provide grounds for accepting it. In both kinds of cases, if we have not irresponsibly ignored clues or hints about how to specify a way in which the statement may be false, we are epistemically entitled to accept it.

# Afterword

As I noted in the Introduction, these essays are efforts to clarify and develop Quine's and Putnam's central insight that we can reject first philosophy without relying on Carnapian linguistic frameworks, by taking scientific method, loosely characterized, yet concretely applied in our best scientific judgments, as our ultimate arbiter of truth, and drawing on our best current judgments, vocabulary, and methods to clarify and facilitate our enquiries. In this final synthesizing section, I briefly trace the arguments and conclusions of the essays back to their deepest source: a new minimalist conception of language use and justification that combines Tarski-style disquotational accounts of truth and satisfaction (reference) for the words we can directly use, with our practical identifications of words between speakers and across time, to yield a minimalist account of transtheoretical terms.

Chapters 1, 2, and 4 explain Carnap's logic of science and Quine's naturalist explication of it, which makes clear that one can reject first philosophy without relying on Carnapian linguistic frameworks. Quine's explication of the logic of science presupposes his minimalist view of language use, according to which we can use our sentences to make assertions, and to support, rescind, or revise our assertions, without implying or presupposing that the meanings of our sentences are fixed by semantical rules. On this view, explicated and defended in Chapter 5, our use of language to make assertions is both pragmatically indispensable to scientific theorizing and independent of particular claims or assumptions about the semantic relations between words and things.

As Chapter 5 also explains, this minimalist elucidation of language use is of a piece with the observation that to get on with one's inquiries, one needs a practical grasp of when it is appropriate to adopt a theory and what it is to justify, defend, and revise one's assertions from the standpoint of a given theory. One attains such a grasp only by immersing oneself, by study and practice, in the relevant scientific disciplines. If, as Quine and

Putnam recommend, one regards scientific method, loosely characterized, yet concretely applied in one's best scientific judgments, as one's ultimate arbiter of truth, one should not seek, in addition, a fully general, discipline-independent account of what it is for one's acceptance of a given theory or statement in any of one's disciplines to be justified.

This minimalist conception of language use and justification is the starting point for my new approach to clarifying and facilitating rational inquiry. The minimalist conception is compatible with introducing regimentations of language to clarify deductive implications and ontological commitments, or facilitate the drawing of inferences, so long as such regimentations are viewed as continuous with all scientific theorizing, in principle revisable, and hence not analytic. Starting with Quine's minimalist conception of language use and justification, for example, we can define satisfaction and truth in Tarski's way for sentences and terms we can directly use. Such notions are better suited to language regimentation, including the kind of regimentations that are needed to formulate logical laws and definitions, than any of our commonsense or intuitive philosophical notions of meaning. For Tarski-style definitions of satisfaction and truth defined for sentences and terms of our own language are as clear to us as our uses of those sentences and terms. They are therefore ideally suited to defining logical truth for sentences that we can use, and, when supplemented by definitional abbreviations, as explained in Chapter 6, to introducing trivial synonymies. Moreover, contrary to Strawson's criticisms of Quine's efforts to define logical truth extensionally, as Chapter 7 shows, we can define logical truth purely extensionally for regimented languages that we introduce by stipulating a Tarski-style truth theory for them. Such stipulations are true by convention in the thin explanatory sense explained in Chapter 3.

Like Quine's definition of logical truth, however, the definition of logical truth developed in Chapter 7 applies in first instance only to sentences of one's present idiolect. It does not show us how to identify cases in which a sentence that one can directly use is logically inconsistent with a sentence of another speaker's idiolect, or of one's own idiolect at some previous time. According to Quine, such applications can be made only relative to a translation from the other speaker's idiolect, or from one's own idiolect at some previous time, into one's present idiolect.

Against this, Putnam stresses that our identifications of transtheoretical terms – i.e., terms whose references are the same despite fundamental differences in the beliefs we and other speakers use the terms to express – are integral to our understanding of truth, which should be understood in a way that fits with our actual practices of identifying agreements and disagreements, adjudicating disputes, and revising our theories. He claims that Quine's Tarski-style definitions of truth, which apply in the first instance only to idiolects, and do not build in our practical identifications of transtheoretical terms, are incoherent.

On the view sketched in Chapter 10, Putnam is right to emphasize that our identifications of transtheoretical terms are integral to our actual practices of identifying agreements and disagreements, adjudicating disputes, and revising our theories, but wrong to claim that Quine's alternative, which does not take such terms as fundamental, is incoherent. (The latter claim begs the question against the minimalist conception of language use that I elucidate in Chapter 5.) Instead, on the reconstruction I recommend, what Putnam's criticisms of Quine's account of truth and reference suggest is that an account of truth and reference is satisfactory only if it fits with and makes sense of our actual practices of agreements and disagreements, adjudicating disputes, and revising our theories. Putnam highlights cases in which scientists now reject a theoretical statement (for example, "Physical space is Euclidean") that they once regarded as beyond doubt, yet continue to take the terms that they first used to assert the statement (including "Physical space") as, in effect, the same, with the same references, as the terms they now use to express their rejection of the statement. As Chapter 10 explains, we can accommodate such cases by combining our practical identifications of words across time and between speakers with Tarski-style definitions of satisfaction for words we can directly use. As Chapter 9 explains, some radical belief revisions lead us to abandon a Bayesean confirmational commitment – i.e., a commitment that settles how we are to update our beliefs in light of new evidence – without leading us to suspend our practical identifications of words across time, or, to put it in terms Quine would prefer, our homophonic translations of previous utterances. This is not to say that all belief revisions are of this kind. As Chapter 8 explains, some belief revisions involve such fundamental shifts in theory that they lead us to reject practical identifications of words, or homophonic translations, for some of our previous utterances. Even in these extreme cases, however, as Chapter 8 argues, there is no reason to conclude that any of sentences we previously accepted are analytic.

Finally, as Putnam observes, for some statements S that we accept, we are unable to specify any way in which S may actually be false. Chapter 11 argues that all we can legitimately say about such statements from our own first-personal methodological standpoint as inquirers is that we are entitled to accept them and hold them immune to disconfirmation unless and until we develop ways to describe how they may actually be false.

# Bibliography

Some of the essays in this volume cite other essays that appear in the volume. The identifying information for these papers is listed in the Acknowledgments section on page iii, and is not also included in the following list.

Alspector-Kelly, Marc, 2001. "On Quine on Carnap on Ontology," *Philosophical Studies* Vol. 102: 93–122.

Austin, J. L. 1946. "Other Minds," *Proceedings of the Aristotelian Society, Supplementary Volumes* Vol. 20: 148–187; reprinted in J. L. Austin, *Philosophical Papers*, 3rd ed., J. O. Urmson and G. J. Warnock (eds.) (Oxford: Oxford University Press, 1979), pp. 76–116.

Auxier, R. E., and Hahn, L. E., 2007. *The Philosophy of Michael Dummett*. La Salle, IL: Open Court.

Baldwin, Thomas, 2006. "Philosophy of Language in the Twentieth Century," in Ernie Lepore and Barry Smith (eds.), *The Oxford Handbook of Philosophy of Language* (Oxford: Oxford University Press 1996), pp. 60–99.

Barrett, Robert, and Gibson, Roger (eds.), 1990. *Perspectives on Quine*. Oxford: Blackwell.

Beaney, Michael (ed.), 1997. *The Frege Reader*. Oxford: Blackwell.

Becker, Edward. 2012. *The Themes of Quine's Philosophy: Meaning, Reference, and Knowledge*. Cambridge: Cambridge University Press.

2006. *Conventionalism*. Cambridge: Cambridge University Press.

Bell, David and Vossenkuhl, Willhelm (eds.), 1992. *Science and Subjectivity: The Vienna Circle and Twentieth Century Philosophy*. Berlin: Akademie.

Benacerraf, Paul. 1973. "Mathematical Truth," *Journal of Philosophy* Vol. 70, No. 19: pp. 661–679.

Ben-Menahem, Yemima. 2005. "Black, White and Gray: Quine on Conventionalism," *Synthese* Vol. 146: 245–282.

Beth, Evert Willem. 1963. "Carnap's Views on the Advantages of Constructed Systems over Natural Languages in the Philosophy of Science," in Schilpp 1963, pp. 469–502.

Bird, Graham, 1995. "Internal and External Questions," *Erkenntnis* Vol. 42, No. 1: pp. 41–64.

Boër, Stephen E. 1977. "Logical Truth and Indeterminacy," *Notre Dame Journal of Formal Logic* Vol. 18, No. 1, pp. 85–94.

Boghossian, Paul. 1996. "Analyticity Reconsidered," *Noûs* 30:3: 360–391.

2000. "Knowledge of Logic," in Boghossian and Peacocke 2000, pp. 229–255.

2001. "How Are Objective Reasons Possible?" *Philosophical Studies* Vol. 106: 1–40.

Boghossian, Paul, and Peacocke, Christopher (eds.), 2000. *New Essays on the A Priori*. Oxford: Oxford University Press.

Bonjour, Laurence, 1998. *In Defense of Pure Reason*. Cambridge: Cambridge University Press.

Boyer, Carl B. 1968. *A History of Mathematics*. Princeton, NJ: Princeton University Press.

Buldt, Bernd. 2004. "On RC 102-43-14," in Steve Awodey and Carsten Klein (eds.), *Carnap Brought Home: The View from Jena* (La Salle, IL: Open Court), pp. 225–246.

Burge, Tyler. 1986. "Intellectual Norms and Foundations of Mind," *Journal of Philosophy* Vol. 83: pp. 697–720.

Carnap, Rudolf. 1928. *The Logical Structure of the World*, 2nd ed., 1967. Berkeley and Los Angeles: University of California Press.

1932. "The Elimination of Metaphysics through Logical Analysis of Language," an English Translation by Arthur Pap of "Überwindung der Metaphysik durch Logische Analyse der Sprache," in *Erkenntnis*, Vol. 2 (1932). The Pap translation is published in A. J. Ayer (ed.), *Logical Positivism* (Glencoe, IL: The Free Press, 1959), pp. 60–81.

1934. "The Task of the Logic of Science," an English translation of a paper originally published in German in 1934, in Brian McGuinness (ed.), *Unified Science: The Vienna Circle Monographs Series Originally Edited by Otto Neurath*. Dordrecht: D. Reidel, 1987, pp. 46–66.

1935a. *Philosophy and Logical Syntax*. Bristol: Thoemmes Press, 1996.

1935b. "Formal and Factual Science." First published in German in *Erkenntnis* 5 (1935); translated and reprinted in Herbert Feigl and May Brodbeck (eds.) *Readings in the Philosophy of Science* (New York: Appleton-Century-Crofts, 1953), pp. 123–128.

1936–1937. "Testability and Meaning," Parts I and II, from *Philosophy of Science* Vol. III, 1936, pp. 419–471, and Vol. IV, 1937, pp. 1–40.

1937. *The Logical Syntax of Language*. An expanded version of what was originally published in German in 1934. Translated by A. Smeaton. London: Kegan Paul, Trench, Trubner & Co.

1939. "Foundations of Logic and Mathematics," *Encyclopedia of Unified Science* Vol. 1, No. 3, Chicago: University of Chicago Press.

1942. *Introduction to Semantics*. Cambridge, MA.: Harvard University Press.

1947. *Meaning and Necessity*. Chicago: University of Chicago Press.

1950. "Empiricism, Semantics, and Ontology," *Revue Internationale de Philosophie* Vol. 4: 20–40; reprinted, with changes, in Carnap 1956, pp. 205–221.

1952a. "Quine on Analyticity," in Creath 1990, pp. 427–432. A transcription and translation of item 102-62-04 in the Rudolf Carnap Collection at the University of Pittsburgh. The manuscript, dated February 3, 1952, is in German shorthand and was transcribed in German by Richard Nollan and translated into English by Richard Creath.

1952b. "Meaning Postulates," *Philosophical Studies* Vol. 3: 65–73; reprinted in Carnap 1956, pp. 222–229

1953.

1956. *Meaning and Necessity*, 2nd ed. Chicago: University of Chicago Press.

1958. *Introduction to Symbolic Logic and its Applications*. Toronto: Dover.

1963a. "Intellectual Autobiography," in Schilpp 1963, pp. 3–84.

1963b. "E. W. Beth on Constructed Language Systems," in Schilpp 1963, pp. 927–933.

1963c. "W. V. Quine on Logical Truth," in Schilpp 1963, pp. 915–922.

Carroll, Lewis 1895. 'What the Tortoise Said to Achilles', *Mind* Vol. 4, No. 14: 278–280.

Chalmers, David 2011."Revisability and Conceptual Change in 'Two Dogmas of Empiricism'," *Journal of Philosophy* Vol. 108, No. 8: 387–415.

Coffa, J. Alberto. 1991. *The Semantic Tradition from Kant to Carnap: To the Vienna Station*. Cambridge: Cambridge University Press.

Creath, Richard. 2007. "Quine's Challenge to Carnap," in Creath and Friedman 2007, pp. 316–335.

1990. *Dear Carnap, Dear Van: the Quine-Carnap Correspondence and Related Work*. Berkeley: University of California Press.

Creath, Richard, and Michael Friedman (eds.), 2007. *The Cambridge Companion to Carnap*. Cambridge: Cambridge University Press.

David, Marian. 1994. *Correspondence and Disquotation: An Essay on the Nature of Truth*. Oxford: Oxford University Press.

Davidson, Donald, 1990. "Meaning, Truth, and Evidence," in Barrett and Gibson 1990, pp. 68–79.

Davidson, Donald, and Hintikaa, Jaakko (eds.), 1969. *Words and Objections*. Dordrecht: D. Reidel.

Deckert, Marion 1973. "Strawson and Logical Truth," *Philosophical Studies* Vol. 24, No. 1: 52–56.

Dummett, Michael, 1978. "The Justification of Deduction," in Michael Dummett, *Truth and Other Enigmas* (Cambridge, MA: Harvard University Press, 1978), 290–318.

1981. *Frege: Philosophy of Language*, 2nd ed. Cambridge, MA: Harvard University Press.

1993. "Wittgenstein on Necessity: Some Reflections," in Michael Dummett, *The Seas of Language* (New York: Oxford University Press, 1993), pp. 446–461.

2007. "Reply to Brian McGuinness," in Auxier and Hahn 2007, pp. 51–54.

Earman, John, 1992. *Bayes or Bust? A Critical Examination of Bayesian Confirmation Theory*, Cambridge, MA: MIT Press.

264 *Bibliography*

Ebbs, Gary. 1992. "Realism and Rational Inquiry," *Philosophical Topics*, vol. 20 (1992): 1–33.

1997. *Rule-Following and Realism*. Cambridge, MA: Harvard University Press.

2000. "The Very Idea of Sameness of Extension across Time," *American Philosophical Quarterly* Vol. 27: 245–268.

2002. "Learning From Others," *Noûs* Vol. 36, No. 4: 525–549.

2003. "A Puzzle about Doubt," in S. Nuccetelli (ed.), *New Essays on Semantic Externalism and Self-Knowledge* (Cambridge, MA: MIT Press), pp. 143–168.

2009. *Truth and Words*. Oxford: Oxford University Press.

2015. "Satisfying Predicates: Kleene's Proof of the Hilbert-Bernays Theorem," *History and Philosophy of Logic* Vol. 36, No. 4: 346–366.

Eklund, Matti, 2013. "Carnap's Metaontology," *Noûs* Vol. 47, No. 2: 229–249.

Etchemendy, John, 1990. *The Concept of Logical Consequence*. Cambridge, MA: Harvard University Press.

2008. "Reflections on Consequence," in Douglas Patterson (ed.), *New Essays on Tarski and Philosophy* (Oxford: Oxford University Press), pp. 263–299.

Enderton, Herbert B. 1972. *A Mathematical Introduction to Logic*. San Diego, CA: Academic Press.

Feferman, Solomon. 1993. "Why a Little Bit Goes a Long Way: Logical Foundations of Scientifically Applicable Mathematics," first published in *PSA 1992*, Vol. 2 (1993): 442–455; reprinted in Solomon Feferman, *In the Light of Logic* (Oxford: Oxford University Press, 1998), pp. 284–298.

Field, Hartry, 1977. "Logic, Meaning, and Conceptual Role," *Journal of Philosophy* Vol. 74: 379–409.

1994. "Deflationist Views of Meaning and Content," *Mind* Vol. 103: 249–285.

Foley, Richard, 1992. "The Epistemology of Belief and the Epistemology of Degrees of Belief," *American Philosophical Quarterly* Vol. 29, No. 2: 111–121.

Frege, Gottlob. 1893. *Grundgesetze Der Arithmetik*, vol. I. Jena: Verlag von Hermann Pohle. Translated into English and edited by Philip A. Ebert and Marcus Rossberg, in *Basic Laws of Arithmetic*, vols. I and II (Oxford: Oxford University Press, 2013).

1902 "Letter to Russell, 22.6.1902. Extract," in Michael Beaney (ed.), *The Frege Reader* (Oxford: Blackwell, 1997), pp. 253–254.

1903. *Grundgesetze Der Arithmetik*, vol. II. Jena: Verlag von Hermann Pohle, 1903. Translated into English and edited by Philip A. Ebert & Marcus Rossberg, in *Basic Laws of Arithmetic*, vols. I and II (Oxford: Oxford University Press, 2013).

1914. "Logic in Mathematics," translated by Peter Long and Roger White, in Hermes, H., Kambartel, F., and Kaulback, F. (eds.), *Gottlob Frege: Posthumous Writings* (Chicago: Chicago University press, 1979), pp. 203–250.

Friedman, Michael. 1988. "Logical Truth and Analyticity in Carnap's *Logical Syntax of Language*," in W. Asprey and P. Kitcher (eds.), *History and Philosophy of Modern Mathematics* (Minneapolis: University of Minnesota Press), 82–94. Reprinted as "Analytic Truth in Carnap's *Logical Syntax of Language*," in Friedman 1999b, pp. 165–176.

1999a. "Tolerance and Analyticity in Carnap's Philosophy of Mathematics," in Friedman 1999b, 198–233.

1999b. *Reconsidering Logical Positivism*, Cambridge: Cambridge University Press.

2007a. "Coordination, Constitution, and Convention: The Evolution of the A Priori in Logical Empiricism," in Alan Richardson and Thomas Uebel (eds.), *The Cambridge Companion to Logical Empiricism* (Cambridge: Cambridge University Press), pp. 91–116.

2007b. "Introduction: Carnap's Revolution in Philosophy," in Creath and Friedman 2007, pp. 1–18.

Frost-Arnold, Greg. 2013. *Carnap, Tarski, and Quine at Harvard: Conversations on Logic, Mathematics, and Science*. Chicago: Open Court.

George, Alexander, 2000. "On Washing the Fur without Wetting It: Quine, Carnap, and Analyticity," *Mind* Vol. 109: pp. 1–24.

Gergonne, Joseph Diaz, 1818. "Essai sur la Théorie des Définitions," *Annales de Mathematiques Pure et Appliquees*, Vol. 9: pp. 1–35.

Giere, Ronald, and Alan Richardson (eds.), 1996. *Origins of Logical Empiricism: Minnesota Studies in the Philosophy of Science*, vol. 16. Minneapolis: University of Minnesota Press.

Gödel, Kurt. 1947. "What is Cantor's Continuum Problem?" First published in *American Mathematical Monthly* Vol. 54: pp. 515–525, 1947; reprinted in Paul Benacerraf and Hilary Putnam (eds.), *Philosophy of Mathematics: Selected Readings* (Cambridge: Cambridge University Press) pp. 470–485.

Gödel, Kurt. 1995. "Is Mathematics Syntax of Language?," Versions III and V, in Solomon Feferman, John W. Dawson, Warren Goldfarb, Charles Parsons, and Robert N. Solovay (eds.), *Kurt Gödel, Collected Works*, vol. III (Oxford: Clarendon Press) pp. 334–362.

Goldfarb, Warren. 1983. "I Want You to Bring Me a Slab: Remarks on the Opening Sections of the *Philosophical Investigations*," *Synthese* Vol. 56, No. 3 (September 1983): 265–282.

1995. Introductory note to Gödel, 'Is Mathematics Syntax of Language?' Versions III and V, Gödel, K. 1995: 324–333.

1996. "The Philosophy of Mathematics in Early Positivism," in Giere and Richardson, 1996, pp. 213–230.

Goldfarb, Warren, and Thomas Ricketts, 1992. "Carnap's Philosophy of Mathematics," in Bell and Vossenkuhl 1992, pp. 61–78.

Gottlieb, Dale. 1974. "Foundations of Logical Theory," *American Philosophical Quarterly* Vol. 11, No. 4: pp. 337–343

Grice, H. P. and Strawson, P. F., 1956. "In Defense of a Dogma," *Philosophical Review* Vol. 65, No. 2: 141–158.

Gupta, A. 1993. "A Critique of Deflationism," *Philosophical Topics* Vol. 21, No. 2: 57–81.

Hacking, Ian, 1990. "Probability," in J. Eatwell, M. Milgate, and P. Newman, *The New Palgrave: Utility and Probability* (New York: Macmillan Press, 1990), pp. 163–177.

Hahn, Lewis E., and Paul A. Schilpp (eds.), 1986. *The Philosophy of W. V. Quine*. La Salle, IL: Open Court.

Harman, Gilbert. 1967. "Quine on Meaning and Existence. I. The Death of Meaning," *Review of Metaphysics* Vol. 21, No. 1: 124–151.

    1974. "Meaning and Semantics," first published in Mildton K. Munitz and Peter K. Unger (eds.), *Semantics and Philosophy* (New York: New York University Pres, 1974), pp. 1–16; reprinted, with changes, in Harman 1999, pp. 192–205

    1976. "Inferential Justification," *Journal of Philosophy* Vol. 73, No. 17: 570–571.

    1986. *Change in View*. Cambridge, MA: MIT Press.

    1994. "Doubts about Conceptual Analysis," first published in M. Michael and J. O'Leary-Hawthorne (eds.), *Philosophy and Mind: The Place of Philosophy in the Study of Mind* (Dordrecht: Kluwer, 1994), pp. 43–48; reprinted, with changes, in Harman 1999, pp. 138–152.

Harman, Gilbert. 1996. Analyticity Regained? *Noûs* Vol. 30: pp. 392–400.

    1999. *Reasoning, Meaning, and Mind*. Oxford: Clarendon Press.

Hylton, Peter. 1982. "Analyticity and Indeterminacy of Translation," *Synthese* Vol. 52: pp. 167–184.

    2007. *Quine*. London: Routledge.

Isaacson, Daniel. 1992. "Carnap, Quine, and Logical Truth," in Bell and Vossenkuhl 1992, pp. 100–130.

Johnsen, Bredo, 2005. "How to Read 'Epistemology Naturalized'," *Journal of Philosophy* Vol. 102, No. 2: 78–93.

Juhl, Cory, and Loomis, Eric. 2010. *Analyticity*. New York: Routledge.

Katz, Jerrold. 1998. *Realistic Rationalism*. Cambridge, MA: MIT Press.

Kim, Jaegwon, 1988. "What Is 'Naturalized Epistemology?'," *Philosophical Perspectives* Vol. 2: 381–405.

Kleene, S. C., 1952. *Introduction to Metamathematics*. Amsterdam, North Holland.

Kripke, Saul, 1976. "Is There a Problem about Substitutional Quantification?" in Evans, G. and McDowell, J. (eds.), *Truth and Meaning: Essays in Semantics* (Oxford: Clarendon Press), pp. 325–417.

    1982. *Wittgenstein on Rules and Private Language: An Elementary Exposition*. Cambridge, MA: Harvard University Press.

Koppelberg, Dirk, 1990. "Why and How to Naturalize Epistemology," in Barrett and Gibson 1990, pp. 200–211.

Lavers, Gregory. 2015. "Carnap, Quine, Quantification and Ontology," in A. Torza, (ed.), *Quantifiers, Quantifiers, and Quantifiers: Themes in Logic, Metaphysics and Language* (Berlin: Springer, 2015), pp. 271–300.

Lepore, Ernie, and Ludwig, Kirk, 2005. *Donald Davidson: Meaning, Truth, Language, and Reality*. New York: Oxford.

Levi, Issac. 1980. *The Enterprise of Knowledge: An Essay on Knowledge, Credal Probability, and Chance*. Cambridge, MA: MIT Press, 1980.

Lewis, David. 1969. *Convention: A Philosophical Study*. Cambridge, MA: Harvard University Press.

Lindley, D. V. 1982. "The Well-Calibrated Bayesian: Comment," *Journal of the American Statistical Association* Vol. 77, No. 379 (1982): 611–612.

Lycan, William. 1991. "Definition in a Quinean World," in J. H. Fetzer, D. Shatz and G. Schlesinger (eds.), *Definitions and Definability: Philosophical Perspectives* (Dordrecht: Kluwer Academic Publishers), pp. 111–131.

Maddy, Penelope. 1997. *Naturalism in Mathematics*. Oxford: Clarendon Press.

2007. *Second Philosophy: A Naturalistic Method*. Oxford: Oxford University Press.

Mancosu, Paulo. 2005. "Harvard 1940–1941: Tarski, Carnap and Quine on a Finitistic Language of Mathematics for science," *History and Philosophy of Logic* Vol. 26: 327–357.

Mancosu, Paulo (ed.), 2008. *The Philosophy of Mathematical Practice*. Oxford: Oxford University Press.

Mates, Benson. 1951 "Analytic Sentences," *Philosophical Review* Vol. 60. 525–534.

McDowell, John, 1984 "Wittgenstein on Following a Rule," *Synthese* Vol. 58 No. 3: 325–363.

McGee, Vann. 1991. *Truth, Vagueness, and Paradox*. Indianapolis: Hackett.

Moore, G. E. 1939. "Proof of an External World." First published in *Proceedings of the British Academy* Vol. 25 (1939): 273–300; reprinted in Thomas Baldwin (ed.), *G. E. Moore: Selected Writings* (London: Routledge, 1993), pp. 147–170.

Morton, Adam. 1973. "Denying the Doctrine and Changing the Subject," *Journal of Philosophy* Vol. 70. 503–510.

Nagel, Thomas, *The Last Word*. New York: Oxford University Press, 1997.

Parsons, Charles. 1995. "Platonism and Mathematical Intuition in Kurt Gödel's Thought," *Bulletin of Symbolic Logic* Vol. 1, No. 1: 44–74.

Peacocke, Christopher, 2000. "Explaining the A Priori: The Programme of Moderate Rationalism," in Boghossian and Peacocke 2000, pp. 255–286.

Potter, Michael. 2002. *Reason's Nearest Kin*. Oxford: Oxford University Press.

Putnam, Hilary. 1962a. "It Ain't Necessarily So," reprinted in Putnam 1975b, pp. 237–249.

1962b. "The Analytic and the Synthetic," reprinted in Putnam 1975c, pp. 33–69.

1971. *Philosophy of Logic*. First published by Harper and Row; reprinted in Hilary Putnam 1975b, pp. 323–357.

1973. "Explanation and Reference," first published in G. Pearce and P. Maynard (eds.), *Conceptual Change* (Dordrecht: Reidel, 1973), pp. 199–221; reprinted in Putnam 1975c, pp. 196–214.

1975a. "The Meaning of 'Meaning'," reprinted in Putnam 1975c, pp. 215–271.

1975b. *Mathematics, Matter, and Method: Philosophical Papers*, vol. 1. Cambridge: Cambridge University Press.

1975c. *Mind, Language, and Reality: Philosophical Papers*, vol. 2. Cambridge: Cambridge University Press.

1975d. "Language and Reality," in Putnam 1975c.

1978. "Reference and Understanding," in Hilary Putnam, *Meaning and the Moral Sciences* (London: Routledge and Kegal Paul), pp. 97–119.

1979. 'Analyticity and Apriority: Beyond Wittgenstein and Quine', first published in P. French et al (eds.), *Midwest Studies in Philosophy*, vol. IV (Minneapolis: University of Minnesota Press); reprinted in Putnam 1983c, pp. 115–138.

1983a. "'Two Dogmas' Revisited," in Putnam 1983c, pp. 87–97.

1983b. "There is at Least One A Priori Truth," in Putnam 1983c, pp. 98–114

1983c. *Realism and Reason: Philosophical Papers*, vol. 3. Cambridge: Cambridge University Press.

1983d. "On Truth", first published in Leigh S. Cauman, Issac Levi, Charles Parsons, and Robert Schwartz (eds.), *How Many Questions? Essays in Honor of Sidney Morgenbesser* (Indianapolis: Hackett); reprinted in Putnam 1994a, pp. 315–329.

1985. "A Comparison of Something With Something Else," first published in *New Literary History* Vol. 17: 61–79; reprinted in Putnam 1994a, pp. 330–350.

1986. "Meaning Holism," in Hahn and Schilpp 1986, pp. 405–426.

1988. *Representation and Reality*. Cambridge, MA: MIT Press.

1991. "Does the Disquotational Theory of Truth Solve All Philosophical Problems?" first published in *Metaphilosophy* Vol. 22, Nos. 1–2: pp. 1–13; reprinted in Putnam 1994a, pp. 264–278.

1993. "The Question of Realism," in Putnam 1994a, pp. 295–312.

1994a. *Words and Life*, edited by James Conant. Cambridge, MA: Harvard University Press.

1994b. "The Dewey Lectures 1994: Sense, Nonsense, and the Senses: An Inquiry into the Powers of the Human Mind," *Journal of Philosophy* Vol. 91, No. 9: pp. 445–517.

1994c. "Rethinking Mathematical Necessity," in Putnam 1994a, pp. 245–263.

Quine, W. V. 1934. "Lectures on Carnap at Harvard University, November 8–22, 1934," in Richard Creath (ed.), *Dear Carnap, Dear Van: The Quine-Carnap Correspondence and Related Work* (Berkeley: University of California Press, 1990), pp. 47–103.

1936. "Truth by Convention," first published in O. H. Lee (ed.), *Philosophical Essays for A. N. Whitehead*. New York: Longmans; reprinted in Quine 1976, pp. 77–106.

1940. *Mathematical Logic*, revised ed. Cambridge, MA: Harvard University Press, 1951. First ed. published in 1940 by Harvard University Press.

1941. *Elementary Logic*. Cambridge, MA: Harvard University Press

1949. "Animadversions on the Notion of Meaning," in W. V. Quine, *Confessions of a Confirmed Extensionalist and Other Essays* (Cambridge, MA: Harvard University Press, 2008), pp. 152–156.

1950. *Methods of Logic*, 1st ed. New York: Henry Holt.

1951. "On Carnap's Views on Ontology," *Philosophical Studies* Vol. 2, No. 5: 65–72.

1952. "On an Application of Tarski's Theory of Truth." First published in *Proceedings of the National Academy of Sciences* Vol. 38: pp. 430–433;

reprinted in W. V. Quine, *Selected Logic Papers*, enlarged ed. (Cambridge, MA: Harvard University Press, 1995), pp. 141–145.

1953a. "On What There Is," *Review of Metaphysics* Vol. 2 (1948): 21–38; reprinted in Quine 1953e, pp. 1–19.

1953b. "Two Dogmas of Empiricism," in Quine 1953e, pp. 20–46.

1953c. "Notes on the Theory of Reference," in Quine 1953e, pp. 130–138.

1953d. "Mr. Strawson on Logical Theory." First published in *Mind* Vol. 62, No. 48 (1953): 433–451, reprinted in Quine 1976, pp. 136–157.

1953e. *From a Logical Point of View.* Cambridge, MA: Harvard University Press.

1959. *Methods of Logic*, revised ed. New York: Holt, Rinehart & Winston.

1960. *Word and Object.* Cambridge, MA: MIT Press.

1963a. "Carnap and Logical Truth," in Schilpp 1963, pp. 385–406.

1963b. *Set Theory and Its Logic.* Cambridge, MA: Harvard University Press.

1969a. "Ontological Relativity," in Quine 1969e, pp. 26–68.

1969b. "Epistemology Naturalized," in Quine 1969e, pp. 69–90.

1969c. "Reply to Chomsky," in Davidson and Hintikaa, 1969, pp. 302–311.

1969d. "Reply to Strawson", in Davidson and Hintikka 1969, pp. 320–325.

1969e. *Ontological Relativity and Other Essays.* New York: Columbia University Press.

1970. *Philosophy of Logic.* Cambridge, MA: Harvard University Press.

1971. "Homage to Carnap." First published in Roger C. Buck and Robert S. Cohen (eds.), *Boston Studies in the Philosophy of Science*, Vol. 8 (Dordrecht: D. Reidel, 1971); reprinted in Creath 1990, pp. 463–466.

1972. *Methods of Logic*, 3rd ed. New York: Holt, Rinehart & Winston.

1974. *The Roots of Reference.* La Salle, IL: Open Court.

1975. "The Nature of Natural Knowledge," in Samuel Guttenplan (ed.), *Mind and Language* (Oxford: Clarendon Press), pp. 67–81.

1976. *The Ways of Paradox*, revised and enlarged ed. Cambridge, MA: Harvard University Press.

1981. "Things and Their Place in Theories," in W. V. Quine, *Theories and Things* (Cambridge, MA: Harvard University Press, 1981), pp. 1–23.

1981c. *Theories and Things.* Cambridge, MA: Harvard University Press.

1982. *Methods of Logic*, 4th ed. Cambridge, MA: Harvard University Press.

1985. *The Time of My Life: An Autobiography.* Cambridge, MA: MIT Press.

1986a. "Intellectual Autobiography," in Hahn and Schilpp 1986, pp. 3–46.

1986b. "Reply to Herbert G. Bohnert," in Hahn and Schilpp 1986, pp. 93–5.

1986c. "Reply to Geoffrey Hellman," in Hahn and Schilpp 1986, pp. 206–208.

1986d. "Reply to Charles Parsons," in Hahn and Schilpp 1986, pp. 396–403.

1986e. "Reply to Putnam," in Hahn and Schilpp 1986, pp. 427–431.

1986f. *Philosophy of Logic*, 2nd ed. Cambridge, MA: Harvard University Press.

1990a. 'Comment on Davidson', in Barrett and Gibson 1990, p. 80.

1990b. "Comment on Koppelberg," in Barrett and Gibson 1990, p. 212.

1991. "Two Dogmas in Retrospect," first published in *Canadian Journal of Philosophy* Vol. 21; reprinted in Quine 2008, pp. 390–400.

1992. *Pursuit of Truth*, rev. ed., Cambridge, MA: Harvard University Press.

1994. "Exchange between Donald Davidson and W. V. Quine following Davidson's lecture," *Theoria* 60: pp. 226–231.

1995. *From Stimulus to Science*. Cambridge, MA: Harvard University Press.

2008. *Confessions of a Confirmed Extensionalist and Other Essays*, edited by Daginn Follesdal and Douglas B. Quine. Cambridge, MA: Harvard University Press.

Reichenbach, Hans, 1958. *The Philosophy of Space and Time*. Maria Reichenbach and John Freund, translators. New York: Dover.

Rey, Georges, 1998. "A Naturalistic A Priori," *Philosophical Studies* Vol. 92 : 25–43.

Richardson, Alan. 1996. "From Epistemology to the Logic of Science," in Giere and Richardson 1996, pp. 309–330.

Ricketts, Thomas. 1982. "Rationality, Translation, and Epistemology Naturalized," *Journal of Philosophy* Vol. 79: pp. 117–136.

Richardson, Alan. 1994. "Carnap's Principle of Tolerance, Empiricism, and Conventionalism," in P. Clark and B. Hale (eds.), *Reading Putnam* (Oxford: Blackwell), pp. 176–200.

1996. "Carnap: from Logical Syntax to Semantics," in Giere and Richardson 1996, pp. 231–250.

2009. "From Tolerance to Reciprocal Containment," in Wagner 2009, pp. 217–235.

1997. "Two Dogmas about Logical Empiricism: Carnap and Quine on Logic, Epistemology, and Empiricism," *Philosophical Topics* Vol. 25: pp. 145–168.

1998. *Carnap's Construction of the World*. Cambridge: Cambridge University Press.

2007. "Carnapian Pragmatism," in Creath and Friedman 2007, pp. 295–315.

Richardson, Alan, and Uebel, Thomas (eds.) 2007. *The Cambridge Companion to Logical Empiricism*. Cambridge: Cambridge University Press.

Romanos, George. 1983. *Quine and Analytic Philosophy*. Cambridge, MA: MIT Press.

Rosen, Gideon. 1999. Review of Penelope Maddy, *Naturalism in Mathematics*. *British Journal of the Philosophy of Science* Vol. 50: pp. 467–474.

Russell, Bertrand. 1903. *The Principles of Mathematics*. Cambridge: Cambridge University Press.

Russell, Gillian. 2014. "Quine on the Analytic/Synthetic Distinction," in Harman, G. and Lepore, E. (eds.), *A Companion to W. V. O. Quine* (Chichester, UK: Wiley-Blackwell), pp. 181–202.

Sarkar, Sahotr. 1992. "'The Boundless Ocean of Unlimited Possibilites': Logic in Carnap's *Logical Syntax of Language*," *Synthese* Vol. 93: pp. 191–237.

Savage, Leonard J. 1967. "Difficulties in the Theory of Personal Probability," *Philosophy of Science* Vol. 34, No. 4: 305–310.

Sayward, Charles 1975. "The Province of Logic," *Analysis* Vol. 36, No. 1, pp. 47–48

Scanlan, Michael. 1991. "Who Were the American Postulate Theorists?" *Journal of Symbolic Logic* Vol. 56: pp. 981–1002.

Schilpp, Paul A. (ed.), 1963. *The Philosophy of Rudolf Carnap*. La Salle, IL: Open Court.

Searle, John R., 1987. "Indeterminacy, Empiricism, and the First Person," *Journal of Philosophy* Vol. 84, No. 3: 123–146.
Seidenfeld, Teddy, Mark J. Schervish, and Joseph B. Kadane, 2012. "What Kind of Uncertainty Is That? Using Personal Probability for Expressing One's Thinking about Logical and Mathematical Propositions," *Journal of Philosophy* Vol. 109, No. 12: 516–533.
Shapere, Dudley. 1969. "Towards a Post-Positivistic Interpretation of Science," in P. Achinstein and S. Barker (eds.), *The Legacy of Logical Positivism* (Baltimore: Johns Hopkins University Press), pp. 115–160.
Sklar, Laurence, 1974. *Space, Time, and Spacetime.* Berkeley: University of California Press.
Shimony, Abner. 1970. "Scientific Inference," in R. G. Colodny (ed.), *The Nature and Function of Scientific Theories* (Pittsburgh: University of Pittsburgh Press), pp. 79–172.
Shoenfield, J. R. 1967. *Mathematical Logic.* Reading, MA: Addison-Wesley.
Soames, Scott. 1984. "What Is a Theory of Truth?" *Journal of Philosophy* Vol. 81: pp. 411–429.
2003. *Philosophical Analysis in the Twentieth Century,* Vol. 1: *The Dawn of Analysis.* Princeton, NJ: Princeton University Press.
2009. "Ontology, Analyticity, and Meaning: the Quine-Carnap Dispute," in David J. Chalmers, David Manley and Ryan Wasserman (eds.), *Metametaphysics: New Essays on the Foundations of Ontology* (Oxford: Oxford University Press, 2009), pp. 424–443.
Stark, Harold. 1995. *Introduction to Number Theory.* Cambridge, MA: MIT Press.
Strawson, P. F. 1950a. "Truth," *Proceedings of the Aristotelian Society, Supplementary Volumes* Vol. 24: 129–156; reprinted in Strawson 1971, pp. 190–213.
1950b. "On Referring," first published in *Mind* Vol. 59, No. 235: 320–344; reprinted in Strawson 1971, pp. 1–27.
1952. *Introduction to Logical Theory.* London: Methuen.
1957. "Propositions, Concepts, and Logical Truths," *Philosophical Quarterly* Vol. 7, no. 26: pp. 15–25; reprinted in Strawson 1971, pp. 116–129.
1971. *Logico-Linguistic Papers.* London: Methuen.
Stroud, Barry. 1969. "Conventionalism and the Indeterminacy of Translation," in Davidson and Hintikaa 1969, pp. 82–96.
1984a. "Skepticism and the Possibility of Knowledge," *Journal of Philosophy* Vol. 81, No. 10: 545–51.
1984b. *The Significance of Philosophical Scepticism.* Oxford: Clarendon Press.
2000. "Quine on Exile and Acquiescence," first published in P. Leonardi and M. Santambrogio (eds.), *On Quine* (Cambridge: Cambridge University Press, 1995), pp. 37–52; reprinted in Barry Stround, *Meaning, Understanding, and Practice* (New York: Oxford, 2000), pp. 151–169.
Talbott, William, 2015. "Bayesian Epistemology," *Stanford Encyclopedia of Philosophy* (Summer 2015 ed.) Edward M. Zalta (ed.), URL: <http://plato.stanford.edu/archives/sum2015/entries/epistemology-bayesian>

Tarski, Alfred. 1936. "The Concept of Truth in Formalized Languages," in Alfred Tarski, *Logic, Semantics, Meta-Mathematics: Papers from 1923 to 1938*, 2nd ed., translated by J. H. Woodger, edited by John Corcoran (Indianapolis: Hackett, 1983), pp. 152–277.

    1941. *Introduction to Logic and to the Methodology of Deductive Sciences*. Oxford University Press.

Thomasson, Amie L. 2015. *Ontology Made Easy*. Oxford: Oxford University Press.

Wagner, Pierre (ed.), 2009. *Carnap's Logical Syntax of Language*. New York: Palgrave Macmillan.

Williams, Michael. 2001. *Problems of Knowledge*, Oxford: Oxford University Press.

Williamson, Timothy. 2007. *The Philosophy of Philosophy*. Oxford: Basil Blackwell.

Wittgenstein, Ludwig. 1921. *Tractatus Logico-Philosophicus*. First published in German in *Annalen der Naturphilosophie*, 1921. Translated by D. F. Pears and B. F. McGuinness. London: Routledge and Kegan Paul, 1961.

    1968. *Philosophical Investigations*, 3rd ed., translated by G. E. M. Anscombe. New York: Macmillan.

    1969. *On Certainty*, edited by G. E. M. Anscombe and G. H. von Wright (Oxford: Basil Backwell, 1969).

Yablo, Stephen, 1998. "Does Semantics Rest on a Mistake?" in *Proceedings of the Aristotelian Society Supplementary* Vol. 72: 229–261.

# Index

agreements/disagreements, 13n3, 28, 30, 50n11, 62, 89n35, 96, 185, 206–207, 211–223, 226, 229, 258–259
  intrasubjective/intersubjective, 13, 62–63, 97, 197, 199, 214, 220n21, 238–239, 251–252, 254
  practical identifications of, 3–4, 31–32, 217, 219–220, 223, 226
Alspector-Kelly, Marc, 33n1
ambiguity, 32, 148, 153, 155–156, 158–159, 225
analytic
  explicated as "confirmed come what may", 178–179, 181–183
  explicated as "L-true", 35, 43–44, 62, 64, 66, 75–77
  and logical consequence, 14–29, 35, 38, 40, 45, 63, 98, 101–103, 107n11
  and logical truth, 6–7, 13, 61–82, 86n33, 92n39, 105, 128–129, 134–141, 146, 169, 174
  Quine's constraints for explicating, 133–136
  statement, 38, 43, 46, 50n11, 79, 128, 134, 136, 169–170, 173, 175, 179, 182, 202
  and transformation rules, 17, 24n14, 45, 51, 54n18, 64, 98–100
  trivially, 4, 6, 33, 35, 38, 42, 45, 48, 49–52, 53n16, 54n17
  understood as "true in virtue of meaning", 34, 135, 145, 177, 191, 224n27
analytic–synthetic distinction, 54, 61–65, 106, 169
analyticity, 7–8, 35, 42, 46, 63, 66, 77, 79, 81, 98–99, 102, 104–106, 128–140, 144–146, 156, 174, 177–179, 182–183, 186, 192, 197, 201–204; see also analytic; analytic–synthetic distinction
a priori
  explicated as "L-true", 34, 54, 73–77
  traditional notion of, 34, 40n7, 53–54, 73–77
a posteriori, 54, 75, 235–237
argument from incomprehension, 183–186

arithmetic, elementary, 15, 18–19, 25–26, 34n3, 35, 63, 71–74, 98–99, 103, 105, 135, 158n3, 244n14
arithmetical, 15–20, 25–26, 74n21
arithmetization, 14–15, 18, 20, 25–26, 103, 109
assertions, 2–3, 5–6, 13, 19, 27, 29, 32, 40n7, 62, 64, 77–78, 96–97, 102, 113–119, 121–127, 156, 162, 184, 205–206, 211–212, 214–215, 220–221, 228–229, 257
Austin, J. L., 145–146, 236n7, 250n17
authority problem for naturalized epistemology, 92–93, 123n2
axioms, 15, 60, 81, 98–99, 236

Baldwin, Thomas, 58
Becker, Edward, 157n2
belief change/revision
  and analyticity, 191–204
  and conceptual role content, 198–201
  and homophonic translation, 168–187, 195–197, 200, 202–203, 259
  intersubjective and intrasubjective, 197, 199
  and translational content, 197, 198n11, 199–200; see also argument from incomprehension; immunity to revision without a change in subject; contextually a priori
Benacerraf, Paul, 58n1
Ben-Menahem, Yemima, ix, 58n1, 79n25, 83n28
Beth, E. W., 31, 75
Bird, Graham, 33n1, 42n8, 52n14
Boër, Stephen E., 153
Boghossian, Paul, 58, 65–66, 68, 128, 130–132, 238, 239n11
Bohnert, Herbert G., 86
Bohr, Niels, 206
Bonjour, Lawrence, 238
Buldt, Bernd, 63
Burge, Tyler, 58

Carroll, Lewis, 59n2
categorical attitudes, 193–194
Chalmers, David, ix, 7–8, 182n7, 191–204
Coffa, J. Alberto, 63
cognitive content, 20–22, 26–29, 32, 38, 41, 44,
    51n13, 53, 62, 99–102, 105, 177
    null, 20–21, 99
    total, 21, 99
conceptual analysis, 2, 54, 119, 125, 127, 163n6
conceptual change/constancy, 191–204
conceptual role, 8, 198
conceptual scheme
    and contextually a priori, 240–251
    as theory, 178n4, 181
    true by virtue of, 173–174, 184
conceptual scheme explanation, 243–251
conditionalization, Bayesian, 8, 191–204
confirmation
    and conceptual side of epistemology, 107–110,
        176–179
    and convention, 108–110
    Quine's minimalist explication of, 110,
        176–177
confirmational commitments
    and Bayesian conditionalization, 198–199, 203
confirmational tenacity, 199, 203
consequence, 14–17, 21, 25–26, 34n3, 35, 45, 59,
    63, 98–99, 102–103, 107
    contrasted with derivability, 15, 35, 72n16, 98
    explicated syntactically, 14–17, 21, 25–26, 35,
        63, 98–99
    explicated semantically, 34n3, 98n3, 147
consistent, 18–19, 23–25, 31, 51 n 13, 67, 71–72,
    74, 80, 87, 105, 212, 214, 244–245, 258
content
    conceptual role, 198–201
    Quine's austere naturalistic view of, 60 n.4,
        169–170, 177, 195, 197, 198n10, 200–201,
        207, 212
    translational, 197, 198n11, 199–200; see also
        cognitive content; empirical content
contextually a posteriori, 235–236
contextually a priori, 224n27, 233–237
contravalid, 16–17, 21, 24n14, 98–100
contradictory, 14, 16–20, 22, 24, 35, 42, 48–49,
    51, 54n17, 63, 70n13, 99
    trivially, 42, 48–49, 51, 54n17; see also consistent
conventionalism
    about logic, 22, 57–78
    about rule-following, 25; see also linguistic
        convention; truth by convention
convention, linguistic. See linguistic convention
Creath, Richard, ix, 50n11, 51n13, 52n15, 58n1,
    72n16, 140n3
credence, 192–195, 198–201, 203, 204

David, Marian, ix, 221n22
Davidson, Donald, 95–96, 108–109, 120n1, 150,
    252n19
Deckert, Marion, 153
decidability, 15, 21
definition, 84n30
    discursive, 89
    legislative, 89, 92, 195; see also definitional
        abbreviations
definitional abbreviations
    convenience of, for logical theory, 138
    and synonymy, 6, 128–146
    as transcriptions of logical truths, 136, 138, 141
demonstrable, 35n4, 158; see also derivable;
    provable
denotation, 207, 210, 217–224, 226–230, 253
derivable, 4, 6, 33, 35, 48–49; see also
    demonstrable; provable
derivation, 16, 35, 63, 155, 161, 244n14
Descartes, René, vi, 114
designation, 42, 46–47, 52n14
determinate, 16–17, 20–22, 24n15, 26, 28–29,
    99–100, 102, 114, 116
diachronic principles of reasoning, 204
discursive postulation. See postulation, discursive
disputes, 3–4, 8, 14, 22, 31–32, 34, 62, 81n27,
    96, 205, 215, 221, 259; see also agreements/
    disagreements
disquotation, 113–116, 119–121, 149–151, 154, 208,
    210–211, 216–229, 257
Duhem, Pierre, 83n28
Dummett, Michael, 25–26, 107, 116, 118, 146,
    235n6

Earman, John, 203n14
Einstein, Albert, 29–30, 170, 234, 241–242
Eklund, Matti, 33n1
elucidation, 123, 125n5, 257
empirical content, 44, 69n4, 169–170, 177, 207, 212
empirical hypothesis, 90, 93
empiricism, 13n14, 33–35, 53, 132–133, 144
    logical, 13n14, 33–35, 53
    and scientific philosophy, 53, 132–133, 144
entitlement to believe that *S*
    contextually a posteriori, 235–236
    contextually a priori, 233–255
epistemology
    and inference, i, 1, 107, 115
    naturalized, 93, 107, 109, 123n2, 212
    and pragmatism, 83n28, 106–107, 163
    and rules for updating beliefs, 203–204
    traditional, 2, 5–7, 13, 34, 40n7, 53–54, 75–77,
        95–96, 101, 103, 106, 108–109, 124–125,
        169, 239; see also diachronic principles of
        reasoning

epistemically reasonable, 8–9, 123, 125, 234, 236, 238, 249

evidence, 5, 8, 86, 93, 95–97, 101, 108–110, 115, 171, 183n9, 192, 194–195, 198–200, 233–234, 241, 243, 249, 259

existence, 4, 33–54

explanation
and postulates, 57, 59–61, 81, 84, 87–88, 92, 94, 186n11
as what counts for most in science, 80–81, 91–92; *see also* laws of logic; P-rules; scientific hypotheses

explication
Carnap's applications of, 27n19, 40n7, 47, 54, 64, 66, 75, 79
method of, contrasted with conceptual analysis, 2–3
Quine's and Quinean applications of, 5, 7, 109, 117–119, 123–127, 133–135, 139–140, 148, 150, 161, 163n6, 169, 177–178, 198–199, 257

external questions, 37–39, 41–42, 48–50

existential instantiation, 45–46

extensionality, 3, 6–7, 133–136, 138–139, 144–167, 169, 258

fact, empirical, 20, 22–24, 26, 28, 74n21

factual, 20, 22, 46, 97–101, 103, 164; *see also* cognitive content

fallibilism, 7, 57, 61, 246, 249–250, 254

Feferman, Solomon, 60

formation rule, 14–16, 21, 34, 36, 64, 98, 101

Field, Hartry, 199n12

fixed-point theorem, Carnap's, 18

framework, linguistic. *See* linguistic framework

Frege, Gottlob, 47n10, 138, 211n11, 244–245

Friedman, Michael, ix, 17n7, 20n21, 54, 58n1, 75, 108–109

Frost-Arnold, Greg, ix, 34n2, 135, 144

fruitless controversies, 22, 28, 30–32, 34, 62, 78, 127; *see also* metaphysics

geometry of physical space, 196, 233n2, 241–242, 245, 252–253

George, Alexander, 81n27

Gergonne, Joseph Diaz, 84n30

Gödel, Kurt, 13, 20n11, 23, 35n4, 63, 74n21

Gödel's first incompleteness theorem, 15, 18–19, 35, 98

Gödel's second incompleteness theorem, 23–24, 72n16, 74n21

Goldfarb, Warren, ix, 13n1, 23, 58n1, 75, 76n22, 118

Goodman, Nelson, 33–35

Gottlieb, Dale, 160n5

Grice, H. P., 6, 105, 116–120, 126, 128–143, 146, 156, 170–172, 181–183, 186

grounds, epistemic, 8–9, 16, 23, 74, 126, 186, 191n1, 200–201, 226–227, 228n33, 234–236, 238, 239n11, 240, 243–245, 251, 254–255

Gupta, Anil, 150

Hacking, Ian, 194n5

Harman, Gilbert, 58n1, 68n10, 69, 70n12, 92n40, 107, 115, 132, 152, 165, 171, 187, 193, 197n8, 198

Holism
and statement verificationism, 118, 177
of theory testing, 82, 106–107, 168n1, 177

Horwich, Paul, x, 216, 221n23

Huntington, E. F., 84n30

Hylton, Peter, ix, x, 60n4, 62, 93–94

idealization of logic, 27–31

idiolect, 123, 142, 208n5, 211n12, 214–216, 219, 223, 258–259

indispensability argument, 60

intensionality, 6, 134, 145, 147, 157–158, 162–166

internal questions, 37–39, 41–42, 52

Isaacson, Daniel, 85n31, 157n2

Jeffrey, Richard, 199n12

Juhl, Cory, 171–172

judgment
context-independent standards for, 81n27, 116, 123–125, 127
and doctrinal side of epistemology, 102, 106–107, 109–110, 114, 123
and justification, 115, 122–123, 162–163
and scientific method, 2, 20, 107, 257–258

justification
and Quine's naturalized epistemology, 5, 106–110, 113, 115–116
in terms of elegance and convenience, 57, 59–61, 81, 88, 93

Kadane, Joseph B., 193n3

Katz, Jerrold, 238, 239n12

Kleene, Stephen Cole, 135, 152

Koppelberg, Dirk, 125n5

Kremer, Michael, x, 244n14

Kripke, Saul, 116–119, 126, 150, 161, 217

Kuhn, Thomas, 203

language system, 14–29, 31, 34–36, 43–44, 47, 49–54, 58, 61–72, 74–78, 82, 87, 97–103, 105–107, 180

language use
and explication of satisfaction and truth, 121
and fallacy of subtraction, 117–118

language use (*cont.*)
  and meaningfulness, 116–117
  Quine's minimal conception of, 3, 5–6, 113,
    117–122, 156, 216–217, 226, 257–259
  Strawson's intensional account of, 145–146,
    148, 156
law-cluster word, 142–143
laws
  logical, 87, 108, 164, 170, 173–174, 207–209,
    211n11, 216–217, 220, 258
  physical, 17, 91, 100, 103, 109–110, 176, 179,
    184–185; *see also* L-rules; P-rules
legislative postulation. *See* postulation,
    legislative
Lepore, Ernie, 120n1
Levi, Isaac, 198–199, 203
Lewis, David, 85n32
liar paradox, 18, 72n17, 208
Lindley, D. V., 193n3
linguistic behavior, 118, 211–212, 216, 222n4
linguistic convention
  Carnap's conception of, 19, 22–25, 28, 57–78,
    97, 100–101, 103–105, 118
  explicit, 6, 19, 22–25, 28, 61–63, 74–76,
    78–79, 97, 100–101, 103–105, 118, 129–132,
    137–138, 140–143
  general, 85
  implicit, 85n32, 92n39
  and meaning, 83
  Quine's conception of, 3, 5, 57, 83–110, 157n2,
    178n4, 258
linguistic framework, 2, 23, 36–37, 39, 43, 45, 47,
    50, 53, 75–76, 78, 257
  customary, 40, 45–48, 52–53
  for speaking about numbers, 37–45
  for speaking about spatiotemporal
    locations, 52–53
  for speaking about things, 39–48
linguistic idealism, 221
Lobachevsky, Nikolai, 233, 241
Lockean Thesis, 193
logic of science, 4–5, 14, 16, 22, 27, 29, 63–64,
    76, 80, 95–110, 257
logical consequence. *See* consequence
logical empiricism. *See* empiricism, logical
logical form, 6, 146–148
logical grammar, 148, 161
logical laws, 87, 91, 176, 211n11
  centrality of, 173–174, 184
  formulation of, using a truth predicate,
    159–160, 207–209, 216–217, 220, 258
  obvious, 108, 174
logical particles. *See* logical words
logical schema, 151–152, 154, 158, 167,
    208–209, 220

logical syntax
  descriptive, 22n12, 36, 80, 101–102
  pure, 66–67, 76, 80, 101–102
logical truth
  and analyticity, 6, 57, 61–78, 128–129, 134–135,
    137–139, 146, 169
  model-theoretic definition of, 151–152
  substitutional definition of, 139, 146, 151–152,
    154–156, 158, 161–163, 167
logical validity, 20, 99–100, 102, 104–105, 146,
    151, 158, 163, 167
logical vocabulary. *See* logical words
logical words, 67, 86
Loomis, Eric, ix, 171–172
Löwenheim–Hilbert–Bernays theorem, 151–152,
    158n3, 167
Ludwig, Kirk, ix, 120n2
Lycan, William, 131, 136

Maddy, Penelope, ix, 60, 88n34
Malcolm, Norman, 203
Mancosu, Paulo, 88n34, 144
McGee, Vann, 210n9
meaning
  and conceptual role, 8, 198nn10–11
  as determined by truth or falsehood of
    sentences, 83
  postulates, 51, 64–65, 97
  Quine's naturalistic explication of, 60n.4,
    169–170, 177, 195, 197, 198n10, 200–201,
    207, 212
  and translation, 62, 86, 101–102, 116–119, 142,
    150, 152, 155, 161, 166, 168, 171–172, 174,
    183–186, 197–200, 207–208, 214–216,
    220–221, 224, 258–259
  truth in virtue of, 16–17, 23, 34, 57–58, 65,
    69–70, 74, 82, 135, 140, 145, 173–174, 177,
    191, 224n27; *see also* content; holism;
    linguistic convention; regimentation;
    synonymy
meaningfulness, 117–119
metalanguage, 26n18, 31–32, 47, 59, 63, 65,
    68–69, 71–72, 74–75, 78, 82, 85, 105,
    149–150, 154, 161–162, 209
metaphysics, 5, 7, 13, 21, 28, 30, 34, 44–45, 62
methodological principles, 1–3, 5, 9, 19–20, 28,
    34, 40n7, 43, 45–46, 50–52, 67n9, 80, 82,
    96–97, 102, 107–108, 110, 123, 124–126,
    178n4, 180, 187, 192, 199–201, 203–204,
    224n27, 229, 252n19, 258
minimal principle of contradiction, 248
Moore, G. E., 227n32, 228n33
Morton, Adam, 89n35
motivating attitude, Carnap's, 13–14, 19, 27,
    29–30, 62

Nagel, Thomas, 126–127
name, 15, 26, 28, 42–48, 50–51, 53n16, 63, 149
naturalism, scientific. *See* scientific naturalism
nominalism, finitistic, 34
normativity, 6, 115

object language, 39–40, 63, 72nn16–17, 74, 150, 161, 209
observational predicate, 26
one-criterion words, 142

paraphrase, 22, 32, 34, 45, 52, 62, 65, 69–72, 74, 119, 126, 148–149, 152–153, 164–166, 168, 179, 194; *see also* explication, regimentation, translation
Peacocke, Christopher, 238
Popper, Karl, 83n28, 100
positivism, 205, 217, 229; *see also* empiricism, logical
postulates
  meaning, 51, 64–65, 97
  of set theory, 57, 59–61, 87–88, 92
postulation
  discursive, 89
  legislative, 89–94, 195
Potter, Michael, 67n9
practical identifications of shared vocabulary and inference rules, 31–32, 220, 257, 259
practical judgment of sameness of denotation, 218–230
pragmatism
  in Carnap's philosophy, 30, 103, 179
  in Quine's philosophy, 83n28, 106–107
Principle of Tolerance, Carnap's, 19–21, 28, 34, 52, 57n9, 102, 103, 107
proposals
  linguistic, 22, 28–29, 32, 40n7, 61, 79, 137, 205
  theoretical, 160–161, 163, 181, 198n11, 220n21, 223n25
propositions, 37n6, 44, 65n7, 68–69, 207–208, 216
protocol sentences, 20–21, 26–27, 49, 99–103, 109
provable, 16, 18–19, 26
pseudoquestion, 41, 50, 53

Quine's observation, 59

realism, 30, 32, 213
reason
  for believing that *S*, 5, 92, 95–96, 106, 115, 186–187, 235, 238–239, 244–255
  faculty of, and apriority, 34–35, 53
reference, as satisfaction, 121, 161, 229n34, 257

regimentation
  and logical truth, 36, 147–153
  Quine's pragmatic method of, 148, 153, 155–156, 158, 160, 162–163
  and synonymy, 148, 150
  and Tarski-style definitions of truth, 147–153, 160–167
regimented language, 3, 6–7, 145, 148–149, 150–167, 209n8, 212n13, 217, 226, 229n34, 254
Reichenbach, Hans, 241
revisability, 2–4, 8, 25, 30, 61, 82, 90–91, 92n40, 106–107, 123n3, 124–126, 134, 150, 162, 168–187, 191–204, 219, 221, 224n27, 228, 241, 248, 250, 255, 257–258
Rey, Georges, ix, 238, 239n12
Riemann, Bernhard, 234, 241
Richardson, Alan, 27n19, 58n1, 80n25, 81n27
Ricketts, Thomas, ix, 13n1, 58n1, 62, 75, 97
Romanos, George, 58
Rorty, Richard, 216, 217n19
Rosen, Gideon, 92–93, 123n2
rule-following, 25, 217n19
Russell, Bertrand, 138, 141, 145, 182, 244–245
Russell, Gillian, 132
Russell's paradox, 244–245
Ryle, Gilbert, 145–146

Sarkar, Sahotra, 15, 18, 63
satisfactionally univocal, 161–163, 167
Savage, Leonard J., 193n3
Sayward, Charles, 160n4
Scanlan, Michael, 84n30
Schervish, Mark, 193n3
scientific
  hypotheses, 79–81, 87, 89–91, 93, 195n7
  judgment, 2, 102, 122–123, 125–127, 186n11, 257–258
  method, 1–3, 5, 13–14, 25–26, 64, 76, 80–81, 93, 96, 104–107, 110, 114–115, 118, 123, 127, 132, 177, 178n4, 192, 203–204, 212–213, 243–259
  naturalism, 6, 57, 59–60, 80, 81n27, 82, 86n33, 88, 91, 93, 104–107, 109, 114, 116, 123–126, 163, 207, 211n11, 220n21, 221n22
  philosophy, 14, 34, 53, 64, 81n27, 119, 123, 123, 132–133, 175
  theory, 5–7, 30, 40n7, 57, 59–60, 79, 81–82, 87–95, 98, 101, 103–104, 106–110, 114–116, 122–127, 152–153, 168, 171, 173, 176–181, 185, 187, 195–196, 205, 206n2, 208, 212, 234, 241–242, 243n13, 253, 257–259
  theory, as field of force, 169, 172, 176, 180
Searle, John, 119–122, 126
Seidenfeld, Teddy, 193n3

Sellars, Wilfred, 208n6
set theory, 15, 20, 51n13, 57, 59–60, 79, 81–82,
    85n31, 87–90, 98, 108–109, 212
Shapere, Dudley, 205n1
Shimony, Abner, 203n14
skepticism, 6, 124–125, 128, 130–131, 143, 235,
    237–240, 244–249, 251, 254–255
Sklar, Lawrence, 242
Significance, 140; see also meaningfulness
Soames, Scott, 33n1, 58, 73–77, 128, 130–132,
    225n29
Strawson, P. F., 6–7, 105, 116–120, 126, 128–132,
    139, 143–150, 153–166, 170–172, 181–183,
    186, 258
Stroud, Barry, ix, 83n28, 95–96, 108, 120n1,
    124n4
substitution, 134–135, 138–139, 151–152, 155–156,
    158, 161–163, 209n7
synonymy
    and confirmation by experience, 182n8
    as created by definitional abbreviations, 6,
        128–143
    different types of, and necessity, 136
    Quine's skepticism about, 6, 116–117, 119, 148,
        156, 165, 207
    and substitutivity salve veritate, 169
    technical notions of, 136; see also
        one-criterion words
syntax. See logical syntax
synthetic, 35, 42, 53

Talbott, William, 203n14
Tarski, Alfred, 33, 71n14, 97, 135, 138, 144, 150n1,
    208; see also Tarski's method of defining
    truth; Tarski's undefinability theorem
Tarski's method of defining truth, 119–120,
    149–151, 154–155, 158–160, 208–210, 224,
    229n34
Tarski's undefinability theorem, 35n4, 71–73,
    78
Thomasson, Amie L., 33n1
translation
    and argument from incomprehension,
        183–186
    homophonic/nonhomophonic, 30, 142, 168,
        171–173, 183–185, 187, 195–196, 197n8, 200,
        202–203, 208n5, 215, 259
    indeterminacy of, 116, 118–121, 166, 207,
        213–220, 229n34
    of logical connectives, 86, 184–185
    manual of, 208, 214, 216

of object language sentences into a
    metalanguage, 150, 155–156, 161, 222–226
Quine's maxim of, 184–186
ultimate parameters for, 219–220
trivial/nontrivial distinction, Carnap's, 38,
    41–42, 48–51, 54
true of, 210n9; see also denotation
true/truth
    as applied to events of utterance, 159–160
    by convention, 1, 3–5, 22–23, 57–94, 105,
        107–108, 129–141, 157n2, 258
    and disquotation, 113–115, 119–121,
        149–151, 154, 162, 208–211, 216–226,
        229n34, 257
    and derivability, 18–19, 35, 63, 98
    predicate, 114, 159, 163, 208, 213, 216–217, 220,
        222–223, 225
    realist semantics of, 213, 216, 229
    and scientific method, 113–114; see also
        analyticity; logical laws, logical truth,
        Tarski's method of defining truth

understanding, 13n3, 49, 62, 86, 96, 102, 110,
    113, 115, 117, 121, 123n3, 125, 183, 205, 212,
    215–216, 218–221, 225, 229, 233, 242–244,
    249–254, 258
universal words, 44–50, 79
univocality, 157, 160; see also weakly univocal;
    satisfactionally univocal
use–mention distinction, 211–212

valid, 16–17, 20–21, 24n14, 29, 98–105, 146, 151,
    158, 163, 167; see also contravalid
variables, 36–38, 42–46, 50–52, 72n17, 137–138,
    149, 209, 253
Vienna Circle, 13n2

weakly univocal, 158–160
Weiner, Joan, ix, 20n11
Williams, Michael, ix, 216, 221n23, 237n10
Wittgenstein, Ludwig, 13, 25–28, 63, 76, 96,
    98, 116, 118, 132–133, 175, 217n19, 227,
    228n33, 233
words
    individuated orthographically, 47, 130,
        159–160
    practical identifications of, 31–32, 215, 218,
        228, 257, 259; see also one criterion words,
        law-cluster word

Yablo, Stephen, ix, 33, 46n9, 53n16